PICTURING MEDICINE
DIFFERENTIAL DIAGNOSES

JOHN K DICKSON
(MB BCHIR MA CANTAB MRCS)

CONTRIBUTORS

MR ALEX VAREY
DR SIMON BULLEY
MISS CHRISTINA UWINS
MR JAMES SMITH
MR AMIT PARMAR
MR JONATHAN KOSY
DR GEORGINA RUSSELL
DR ZUDIN PUTHUCHEARY
DR CAROLINE JOHNSTON
DR THOMAS SANCTUARY
DR JEAN RICHARDSON
DR CHERN-SIANG LEE
MR BEN BYRNE
DR TIM DAWES
DR ALEX DAVIES

Picturing Medicine
Differential Diagnoses

First published 2010

ISBN 978-1-4467-6894-5

www.PicturingMedicine.com

Picturing Medicine
Differential Diagnoses

Preface.

When I was studying at medical school I found it very difficult to assimilate and remember all the information that was thrown my way. I was not alone, and all my fellow student colleagues found the shear volume of imformation very challenging. We studied hard and crammed information into our brains, only to find lots of what we had learnt would fade in our memories within weeks or even days of the exams! In response to these pressures, I developed a visual approach to my studies. I found that it was always the images and visual information that seemed to stay with me. Many people are visual learners and remember more effectively when they use images and drawings.

These visual aide-memoires have been designed to assist medical students and healthcare professionals with rapid recall of facts and clear thinking. The aim has been to produce images which can help to build a strong foundation of core clinical knowledge. The visual aide-memoire can act as an associative "mind-map". The student can read around on the topics and gain extra clinical experience to further build from this strong visual foundation!

Ultimately, the visually associative approach allows the user to recall information rapidly with accuracy and confidence. This will be especially helpful in emergency situations when healthcare professionals are called on to act quickly and to recall information that they may not use on a regular basis.

Knowledge that is required for undergraduate exams is also required for postgraduate exams so enhancing your learning with these visual aides will serve you well in the future too!

PICTURING MEDICINE
DIFFERENTIAL DIAGNOSES

The diagrams have been specifically designed to be visually associative and peculiar. This is intentional as strange images which are visually associative will tend to stick in the mind!

In addition, I have included the images with the labels REMOVED as this focuses the user on the IMAGE.

Do add hand written notes and scribbles if you wish as this will also help you learn!

If you have any ideas, suggestions or feedback then please do contact me.

Best wishes,
John K Dickson
www.PicturingMedicine.com

About the author.

John Dickson is a plastic surgery trainee currently working in South West England.

He studied medicine at the University of Cambridge, qualifying in 2005.

He is a member of the Royal College of Surgeons of England.

PICTURING MEDICINE
DIFFERENTIAL DIAGNOSES

Contents

www.PicturingMedicine.com

PICTURING MEDICINE
DIFFERENTIAL DIAGNOSES

Removing the labels.

Once you understand the diagram, focus on the image with the labels removed!

By focusing on the image WITHOUT the labels you will have a clearer visual picture in your minds eye.

Try it!

Causes of **Abdominal Pain**

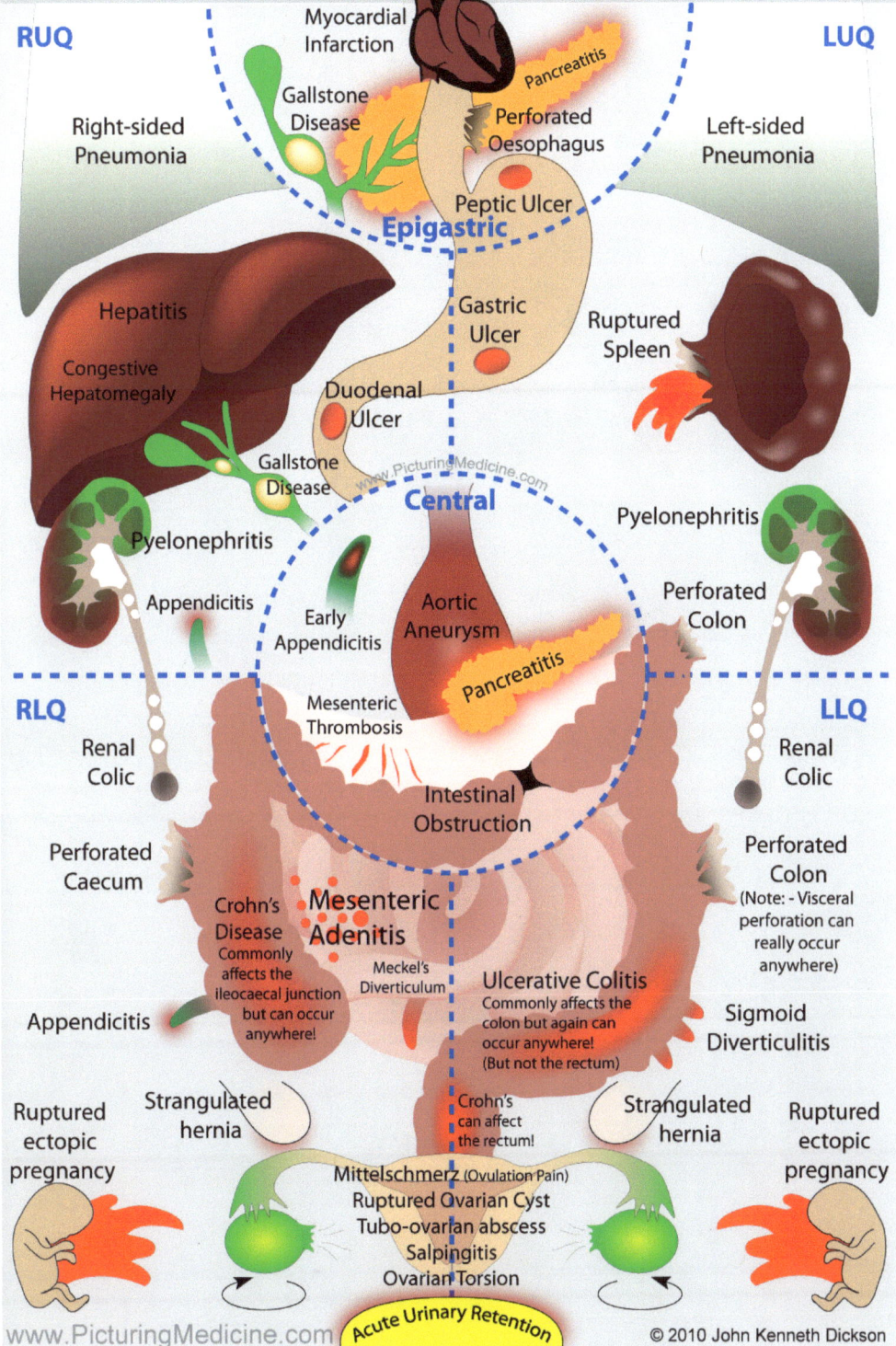

RUQ

LUQ

Myocardial Infarction

Pancreatitis

Gallstone Disease

Perforated Oesophagus

Right-sided Pneumonia

Left-sided Pneumonia

Peptic Ulcer

Epigastric

Hepatitis

Gastric Ulcer

Ruptured Spleen

Congestive Hepatomegaly

Duodenal Ulcer

Gallstone Disease

www.PicturingMedicine.com

Central

Pyelonephritis

Pyelonephritis

Appendicitis

Early Appendicitis

Aortic Aneurysm

Perforated Colon

Pancreatitis

RLQ

Mesenteric Thrombosis

LLQ

Renal Colic

Renal Colic

Intestinal Obstruction

Perforated Caecum

Crohn's Disease
Commonly affects the ileocaecal junction but can occur anywhere!

Mesenteric Adenitis

Meckel's Diverticulum

Perforated Colon
(Note: - Visceral perforation can really occur anywhere)

Appendicitis

Ulcerative Colitis
Commonly affects the colon but again can occur anywhere!
(But not the rectum!)

Sigmoid Diverticulitis

Ruptured ectopic pregnancy

Strangulated hernia

Crohn's can affect the rectum!

Strangulated hernia

Ruptured ectopic pregnancy

Mittelschmerz (Ovulation Pain)
Ruptured Ovarian Cyst
Tubo-ovarian abscess
Salpingitis
Ovarian Torsion

Acute Urinary Retention

© 2010 John Kenneth Dickson

1

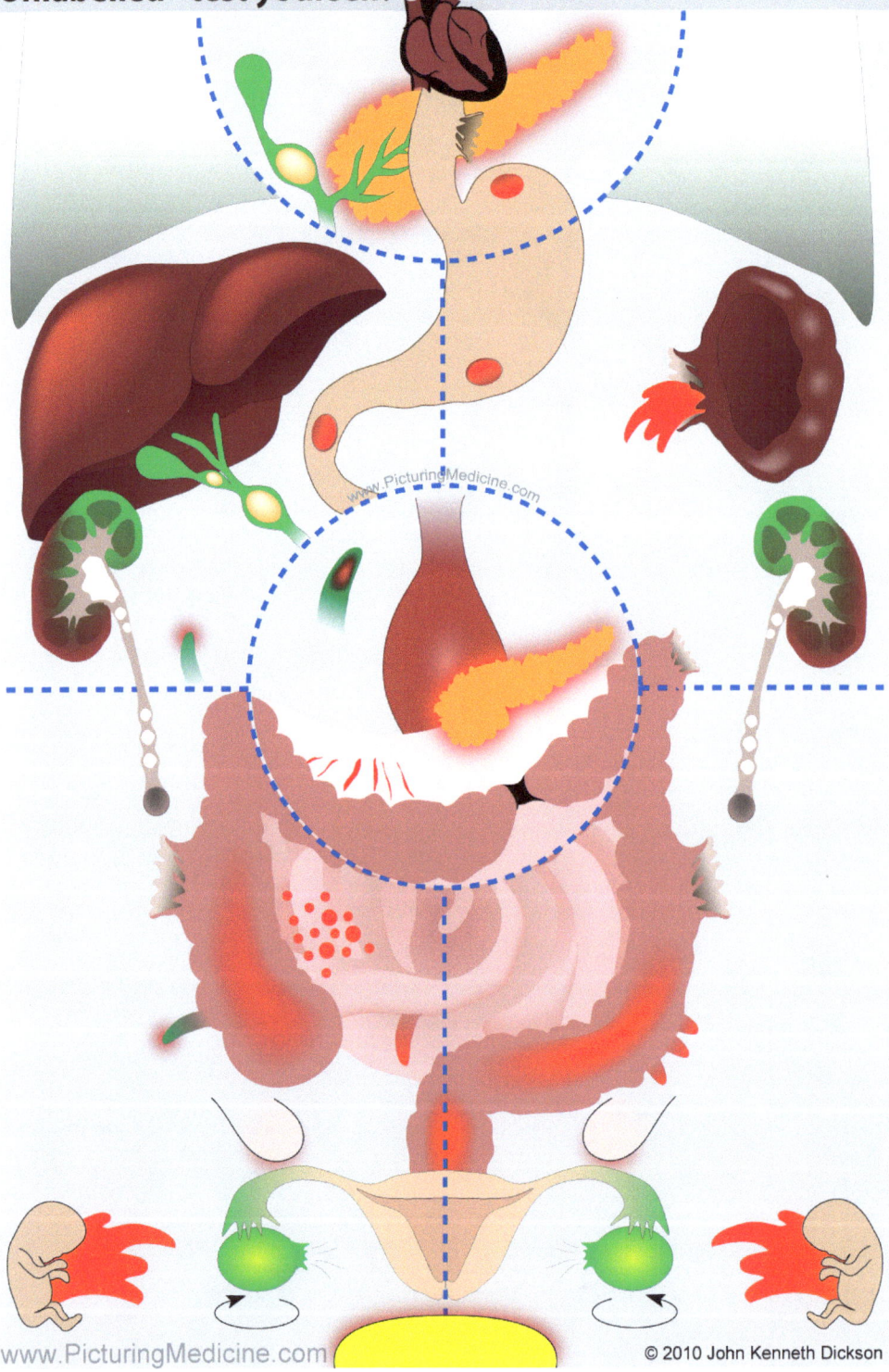

2

The causes of **Shortness of Breath**

Anaphylaxis

Central causes
(leading to
hypo-
ventilation)

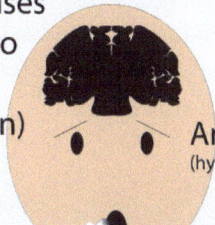

H+ H+ H+ H+ H+ H+ **Metabolic**
H+ H+ **Acidosis** H+
H+ H+ H+
H+ H+
H+

Anxiety
(hyperventilation)

STRIDOR

Upper Airway Obstruction
often with STRIDOR

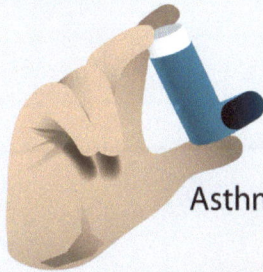

Asthma

Aspirin OD

Laryngeal
Fractures

Pulmonary Fibrosis
"honeycomb lung"

Phrenic Nerve
lesion (C3,4,5)

Pneumo
-thorax

Chronic
Obstructive
Lung Disease

Lung Cancer

Pulmonary
Embolus

Kyphoscoliosis

Pulmonary
Collapse

Bronchiectasis

Basal Atelectasis
Pneumonia

Heart
Failure

Myocardial
Infarction

Pleural
Effusion

www.PicturingMedicine.com

Diaphragmatic Weakness
eg Myasthenia Gravis, Gullain-Barre Syndrome
MS, Polio, MND, Muscular Dystrophy

Arrhythmias

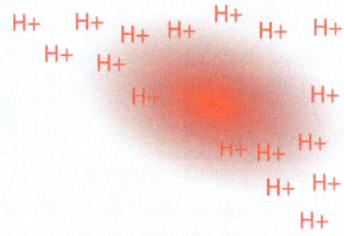

4

The causes of **Chest Pain**

Life-threatening causes ...

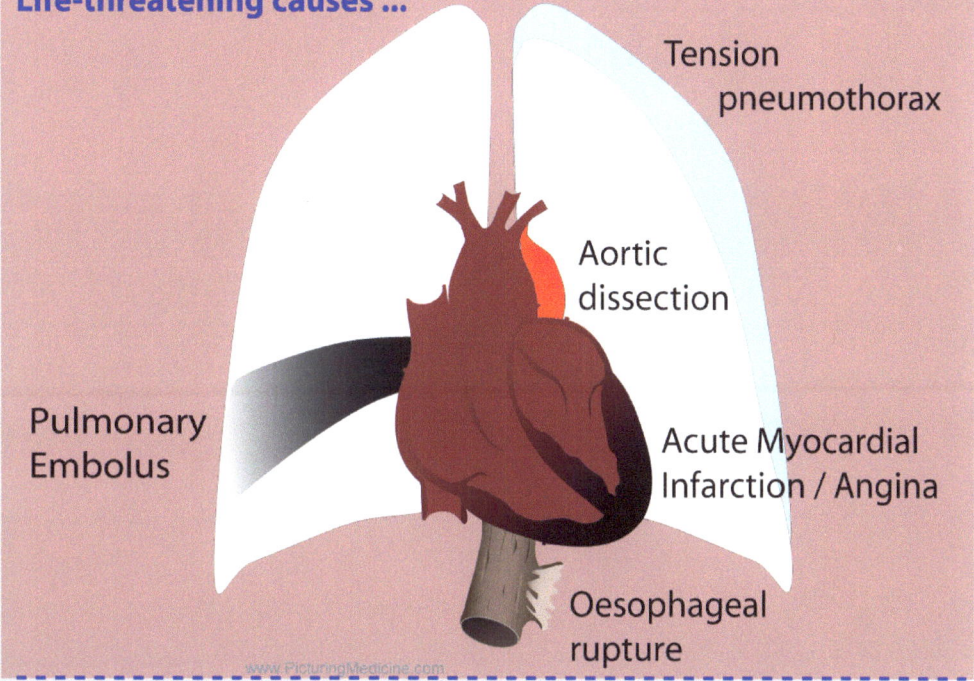

Tension pneumothorax

Aortic dissection

Pulmonary Embolus

Acute Myocardial Infarction / Angina

Oesophageal rupture

www.PicturingMedicine.com

Other causes ...

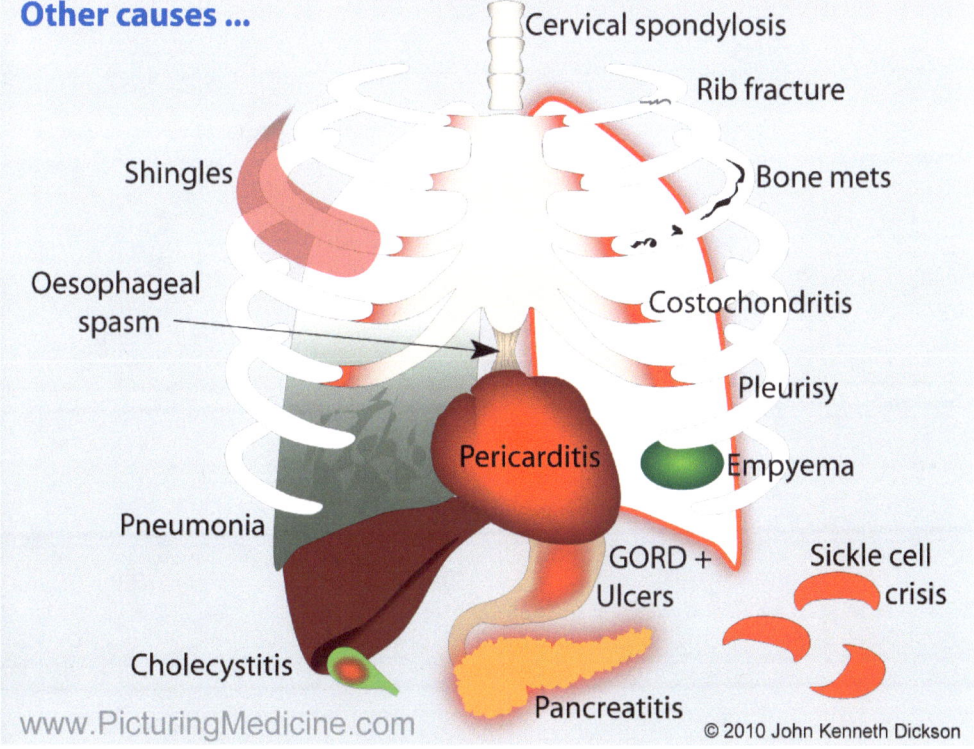

Cervical spondylosis

Rib fracture

Shingles

Bone mets

Oesophageal spasm

Costochondritis

Pleurisy

Pericarditis

Empyema

Pneumonia

GORD + Ulcers

Sickle cell crisis

Cholecystitis

Pancreatitis

www.PicturingMedicine.com

© 2010 John Kenneth Dickson

5

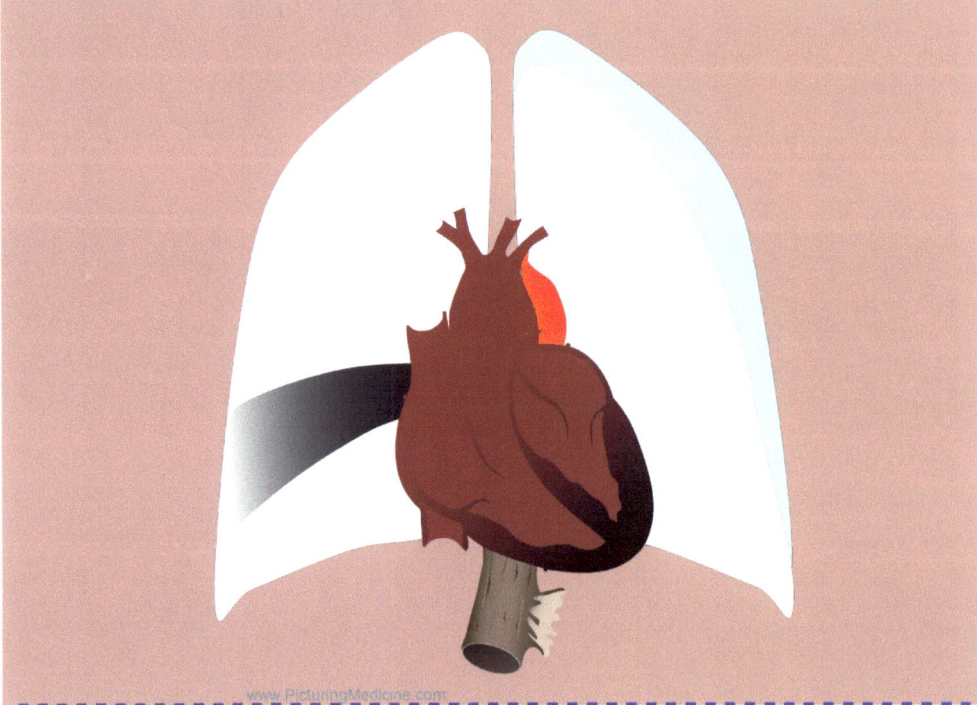

www.PicturingMedicine.com

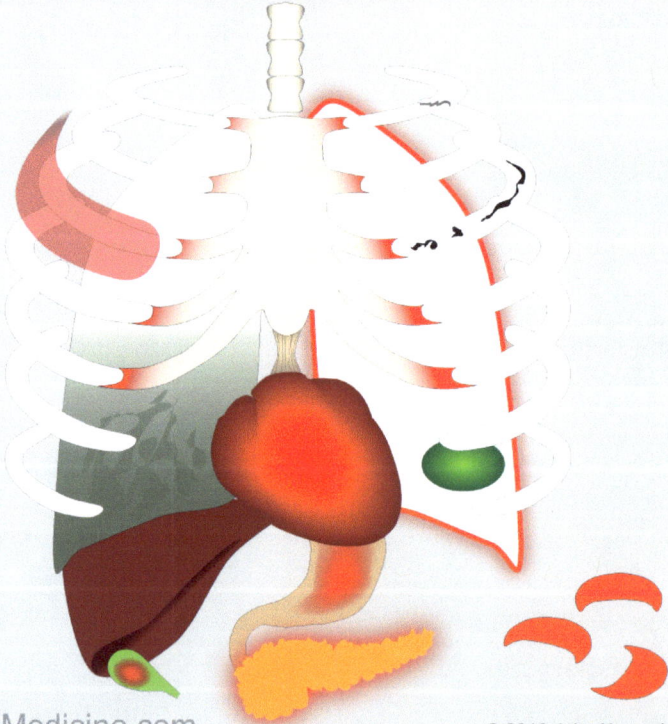

The causes of **Coma** *(unrousable and unresponsive patient)*

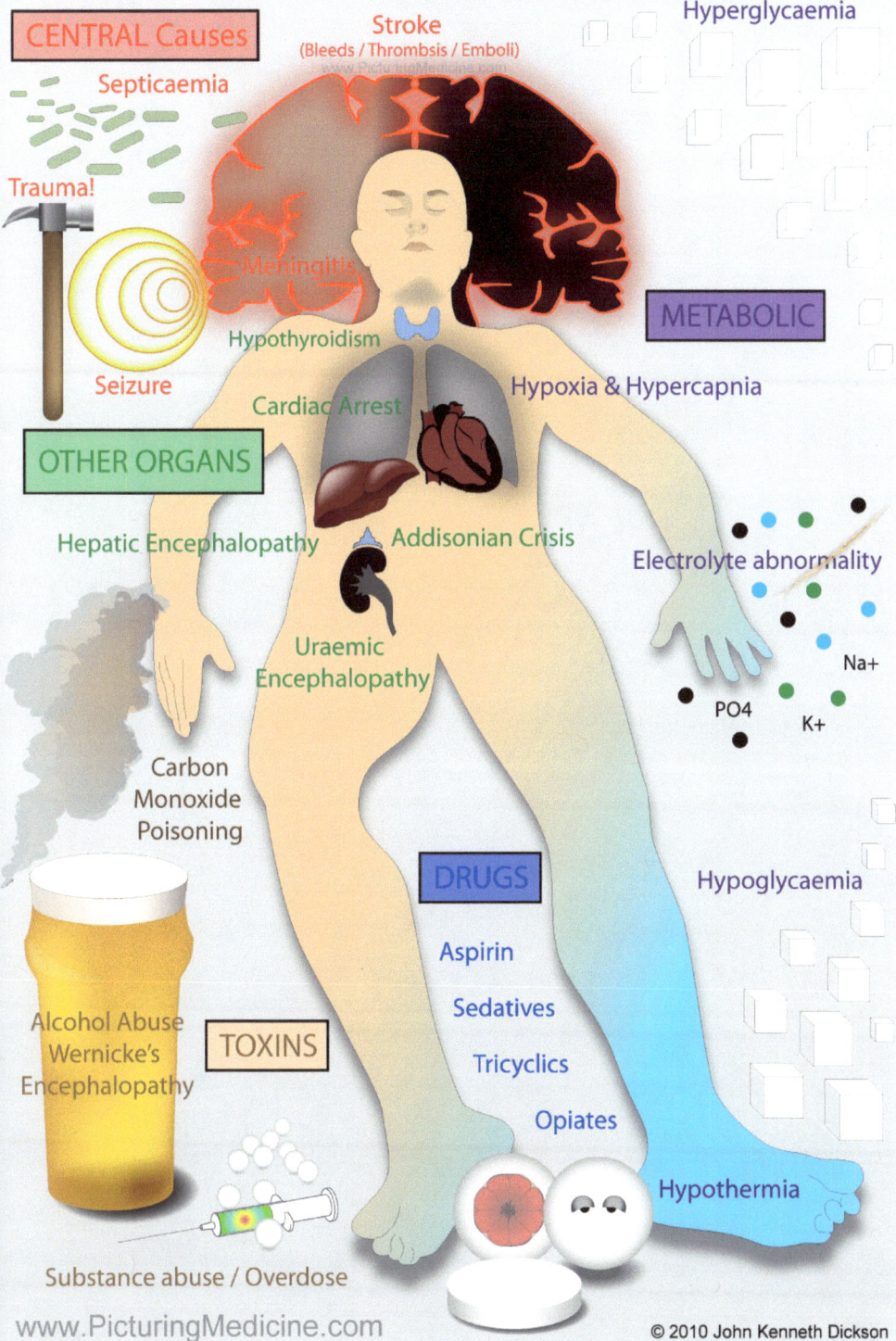

CENTRAL Causes

Septicaemia

Trauma!

Seizure

Stroke
(Bleeds / Thrombsis / Emboli)
www.PicturingMedicine.com

Meningitis

Hypothyroidism

OTHER ORGANS

Cardiac Arrest

Hepatic Encephalopathy

Addisonian Crisis

Uraemic
Encephalopathy

Carbon
Monoxide
Poisoning

Hyperglycaemia

METABOLIC

Hypoxia & Hypercapnia

Electrolyte abnormality

Na+

PO4

K+

Hypoglycaemia

DRUGS

Aspirin

Sedatives

Tricyclics

Opiates

Hypothermia

Alcohol Abuse
Wernicke's
Encephalopathy

TOXINS

Substance abuse / Overdose

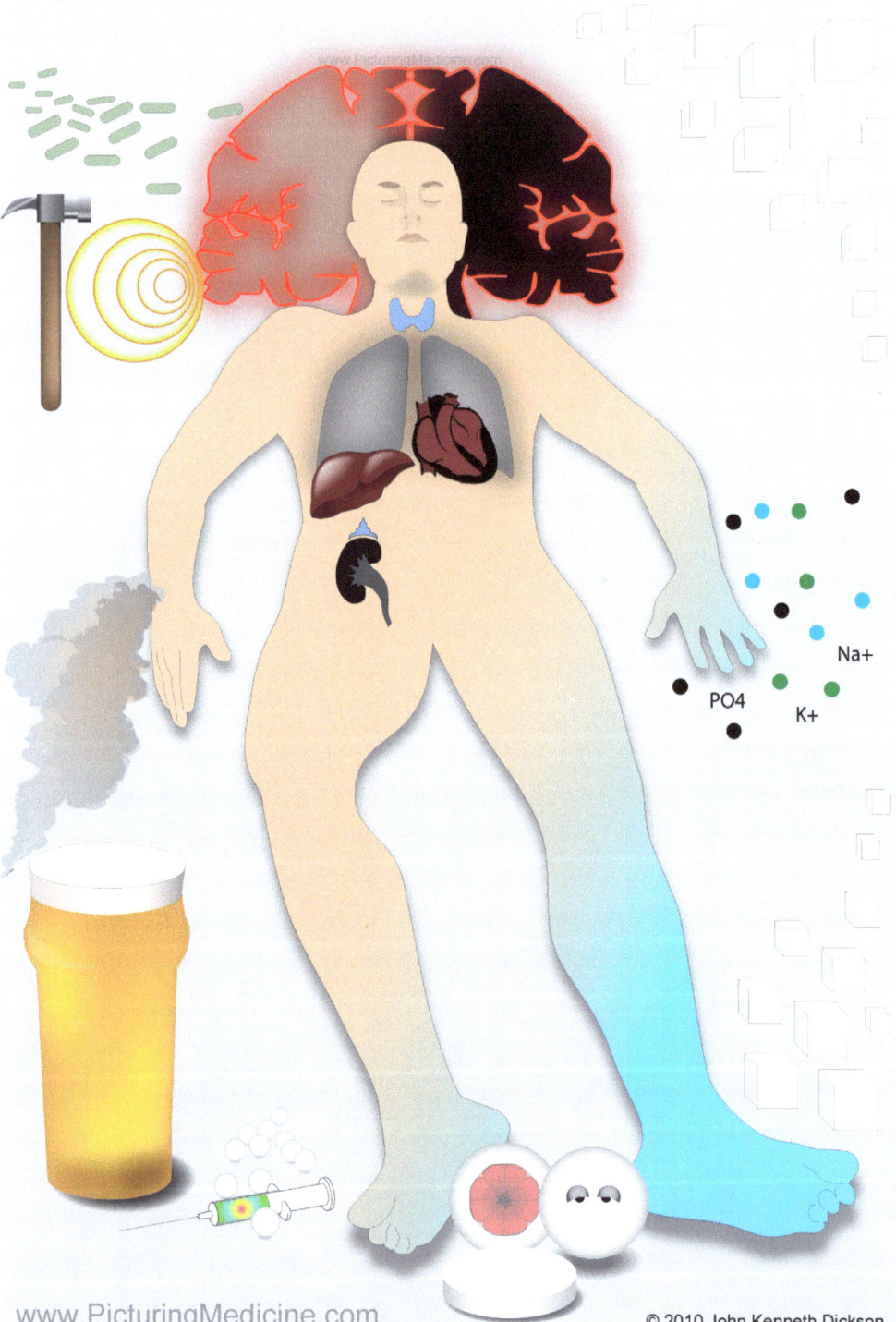

Na+

PO4

K+

8

The causes of **Haemoptysis**

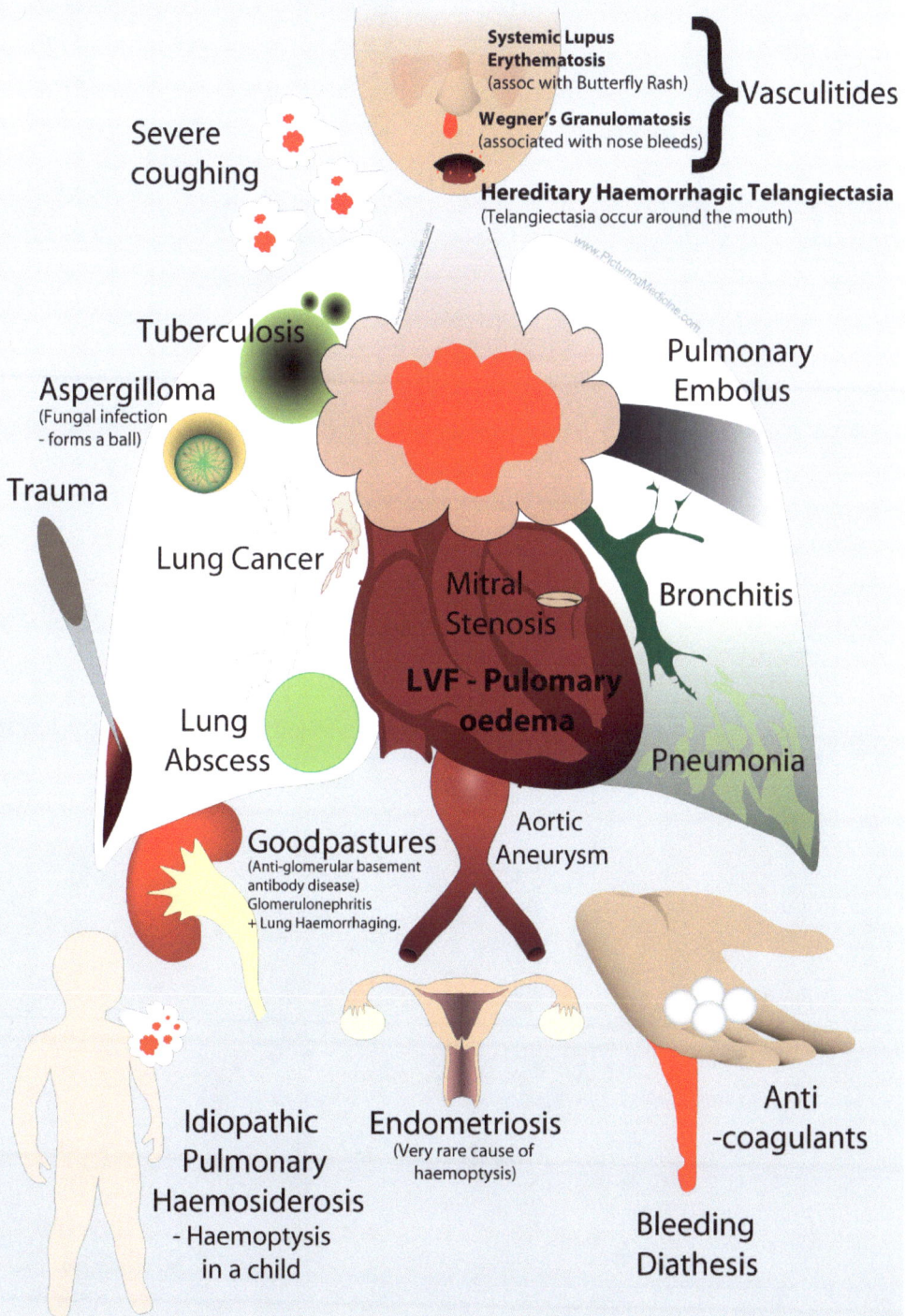

Systemic Lupus Erythematosis (assoc with Butterfly Rash)

Wegner's Granulomatosis (associated with nose bleeds)

} Vasculitides

Hereditary Haemorrhagic Telangiectasia
(Telangiectasia occur around the mouth)

Severe coughing

Tuberculosis

Pulmonary Embolus

Aspergilloma
(Fungal infection - forms a ball)

Trauma

Lung Cancer

Mitral Stenosis

Bronchitis

LVF - Pulomary oedema

Lung Abscess

Pneumonia

Goodpastures
(Anti-glomerular basement antibody disease)
Glomerulonephritis + Lung Haemorrhaging.

Aortic Aneurysm

Anti-coagulants

Idiopathic Pulmonary Haemosiderosis
- Haemoptysis in a child

Endometriosis
(Very rare cause of haemoptysis)

Bleeding Diathesis

9

The causes of **Headache**

Carbon monoxide poisoning

Altitude Sickness

Malaria

Migraine
(Unilateral + throbbing.
Vision can be affected)

Raised Intra-cranial Pressure

Meningism

Giant Cell Arteritis

Tension Headache

Trigeminal Neuralgia
(Pain in the trigeminal nerve distirbution)

Blurred vision!

Cluster Headache

Post-concussion

Sinusitis

Malignant Hypertension

Glaucoma
(Causes blurred vision)

Subarachnoid Haemorrhage
(Sudden, severe headache)

Cervical Spondylosis

Stroke

Drugs
eg Nitrates,
Ca Ch antagonists

Pre-eclampsia
(In third trimester of pregnancy)

The causes of **Per Vaginal Bleeding** (outside of pregnancy)

Bleeding unrelated to menstruation

Causes associated with heavy periods.

Drugs
eg progesterone-only
oral contraceptive pill .

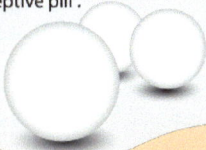

Hypothyroidism
(A rare cause)

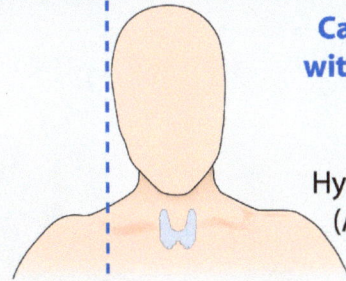

Pelvic
Inflammatory
Disease - *Infection!*

Intra-
uterine
Coil

Fibroids

Uterine
Cancer

Endometrial Hyperplasia

Endometrial Polyps
(Also cause post-menopausal
bleeding)

Cervical Cancer,
Polyps & Erosions

Post-menopausal
Atrophic Vaginitis
(Associated with decreased
oestrogen levels)

Dysfunctional
Uterine Bleeding
(Diagnosis of exclusion.
Affects middle aged women
causing peri-menopausal
menorrhagia)

Vaginal
Cancer

Post-op

Drugs
Anti-coagulants!
(Rarely seen)

Thrombocytopenia

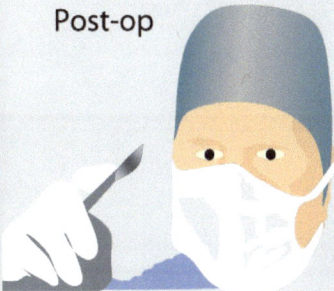

Trauma
- Common cause
of bleeding.

www.PicturingMedicine.com

© 2010 John Kenneth Dickson

www.PicturingMedicine.com

13

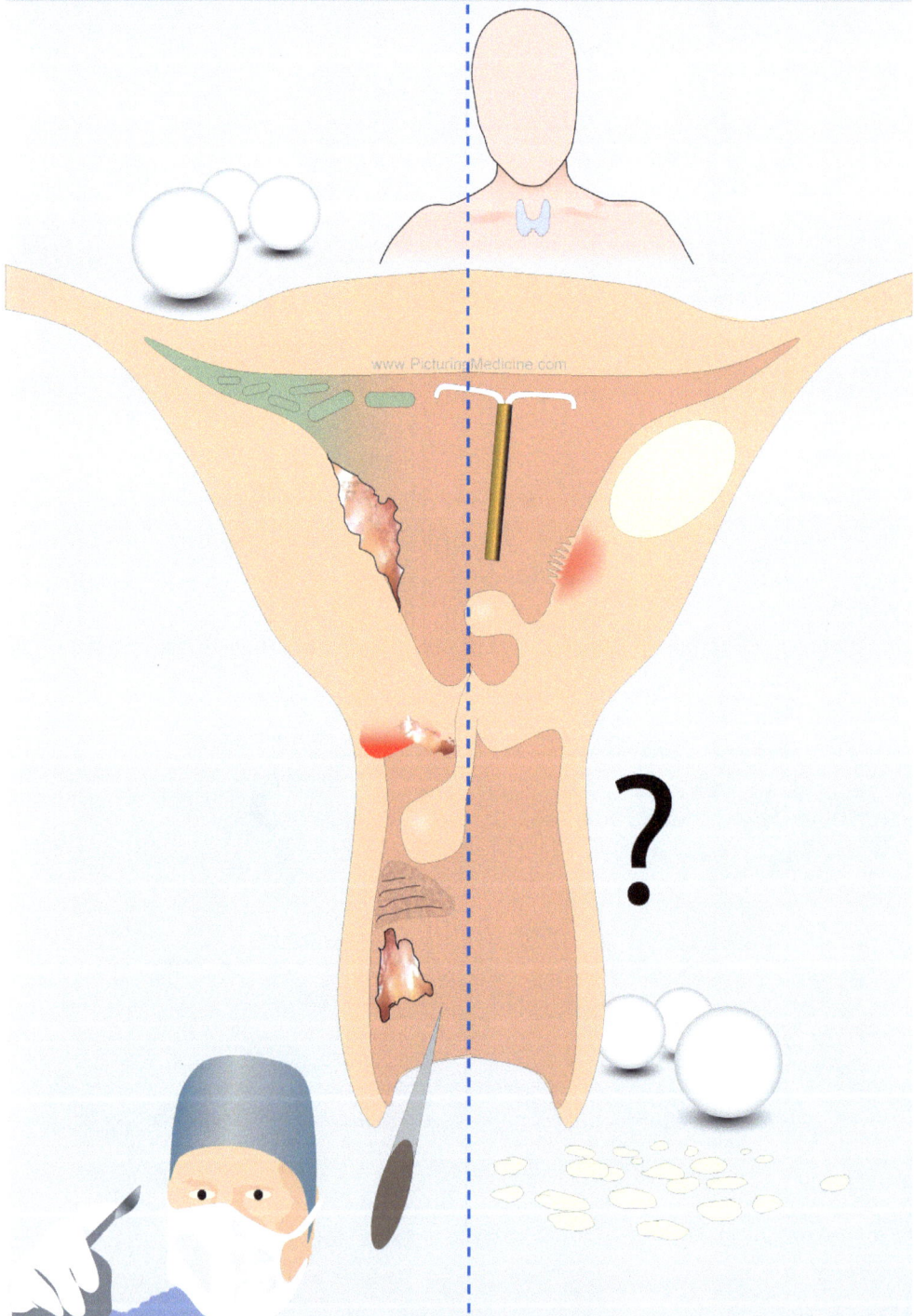

© 2010 John Kenneth Dickson

14

Complex Burns (which require referral to a specialist unit)

Burns are considered to be complex when associated with any of the following...

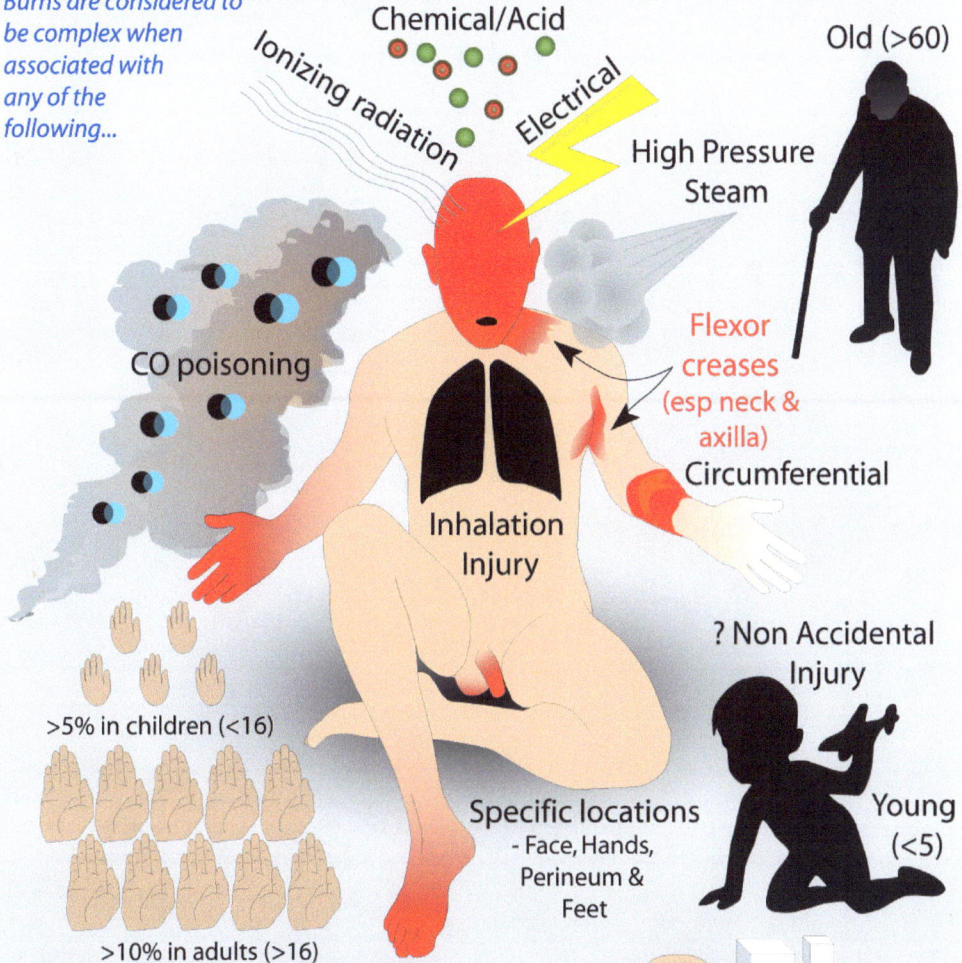

Ionizing radiation

Chemical/Acid

Electrical

High Pressure Steam

Old (>60)

CO poisoning

Inhalation Injury

Flexor creases (esp neck & axilla)

Circumferential

? Non Accidental Injury

>5% in children (<16)

>10% in adults (>16)

Specific locations
- Face, Hands, Perineum & Feet

Young (<5)

A burn of a significant size with **dermal** or **full thickness** loss
(palmar surface of the hand from fingertips to wrist = 1%)...

sugar!

Associated Injuries!
- Head Injry
- Fracture
- Penetrating Injury
- Crush Injury

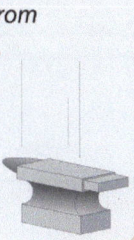

Pre-existing conditions!
- Diabetes
- Cardiac/ Respiratory
- Liver failure
- Pregnancy
- Immunosuppression

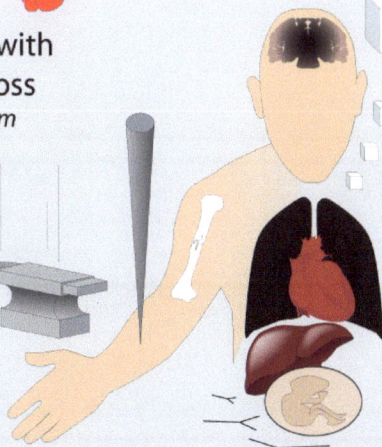

The causes of **Back Pain**

Common causes ...

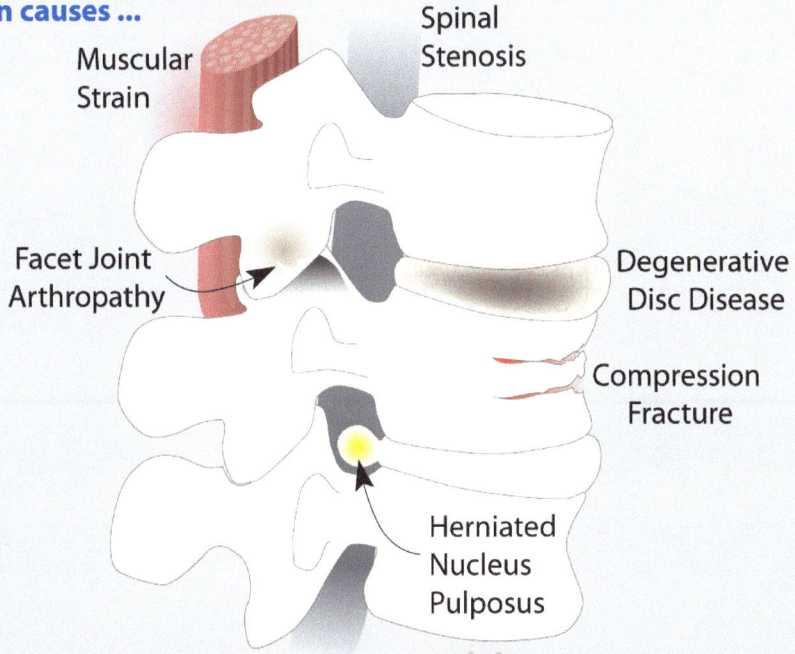

Muscular
Strain

Spinal
Stenosis

Facet Joint
Arthropathy

Degenerative
Disc Disease

Compression
Fracture

Herniated
Nucleus
Pulposus

**Less common
causes ...**

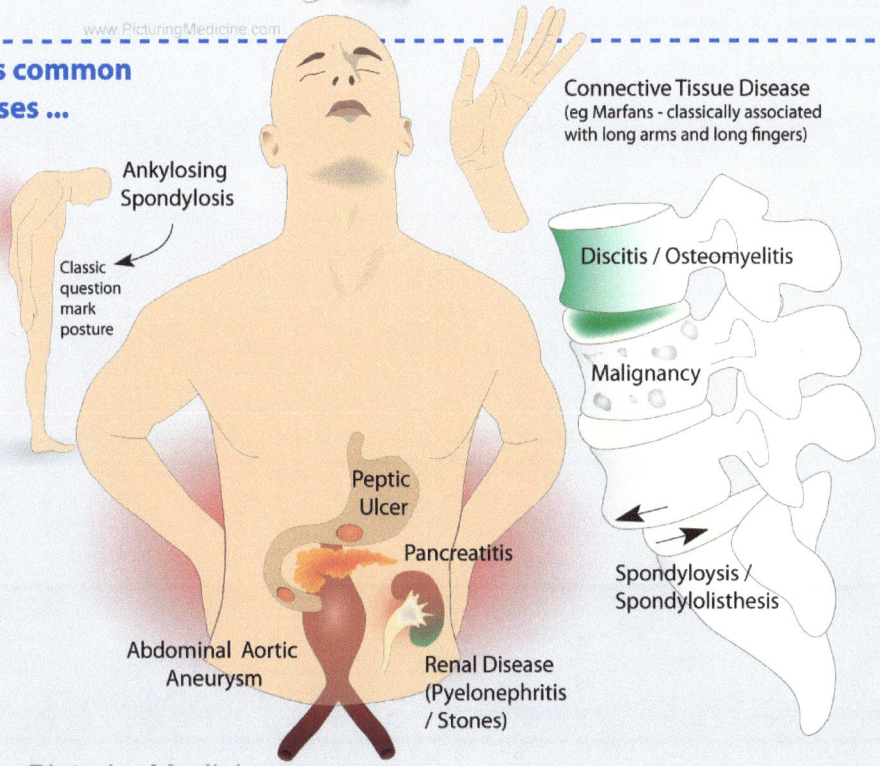

Ankylosing
Spondylosis

Classic
question
mark
posture

Connective Tissue Disease
(eg Marfans - classically associated
with long arms and long fingers)

Discitis / Osteomyelitis

Malignancy

Peptic
Ulcer

Pancreatitis

Abdominal Aortic
Aneurysm

Renal Disease
(Pyelonephritis
/ Stones)

Spondyloysis /
Spondylolisthesis

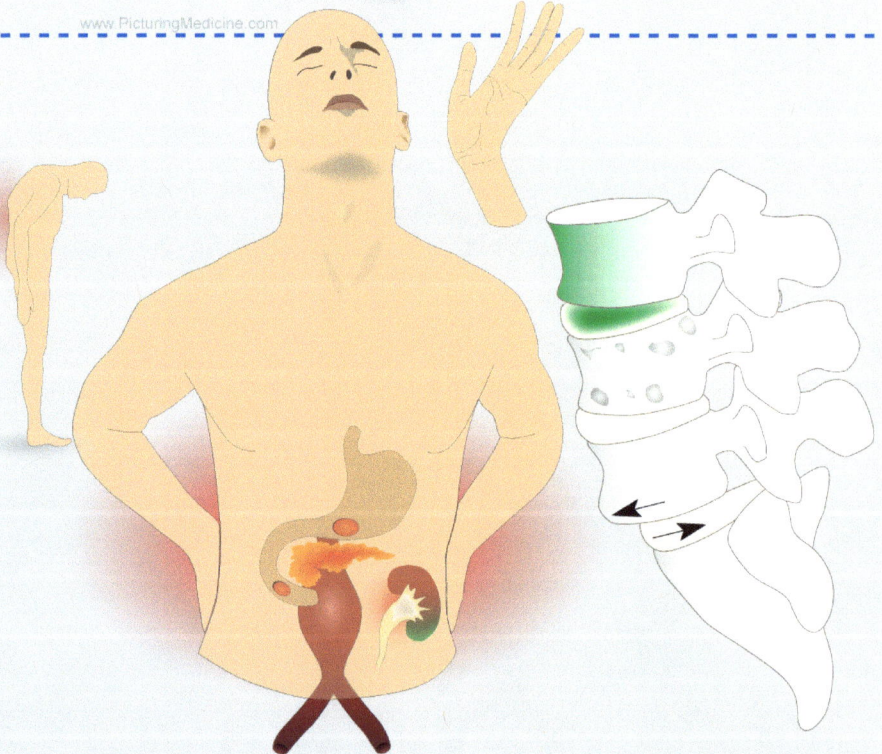

18

The causes of **Abdominal Pain in a Child**

Surgical Causes

Medical Causes

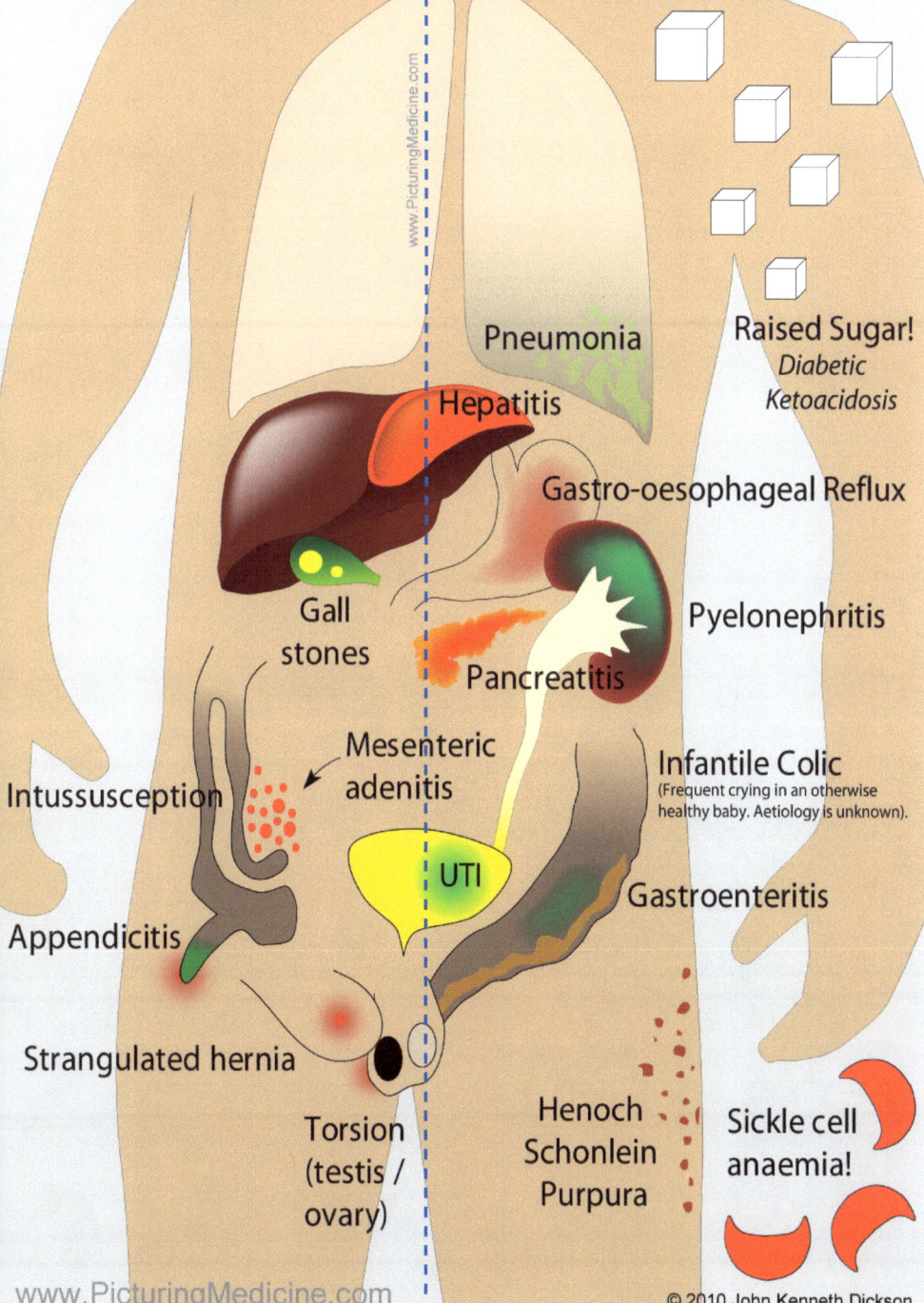

Pneumonia

Hepatitis

Raised Sugar!
Diabetic Ketoacidosis

Gastro-oesophageal Reflux

Gall stones

Pyelonephritis

Pancreatitis

Mesenteric adenitis

Infantile Colic
(Frequent crying in an otherwise healthy baby. Aetiology is unknown).

Intussusception

UTI

Gastroenteritis

Appendicitis

Strangulated hernia

Torsion (testis / ovary)

Henoch Schonlein Purpura

Sickle cell anaemia!

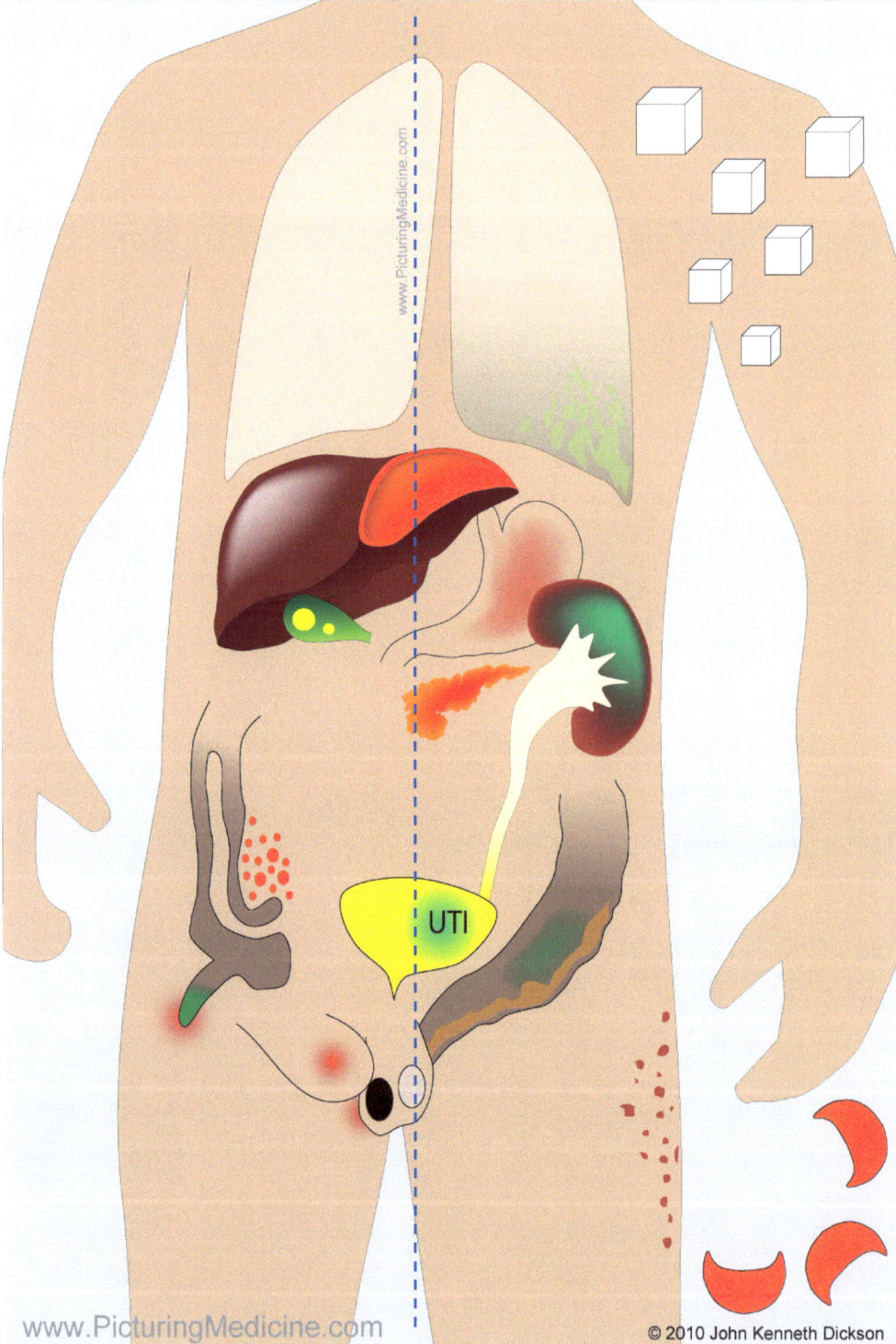

UTI

The causes of **Fever in a Child**

Common causes

Less common causes

Non-specific
Viral
Infection
(common)

Malaria (Transmitted by moquitos)

Familial
Mediterranean
Fever

Meningitis
(Potentially life
threatening!)

Otitis media
(Ear infection)

Tonsillitis

Systemic Lupus Erythematosis
(A form of vasculitis classically associated
with a butterfly rash over the cheeks)

Tuberculosis

Kawasaki
Disease

Childhood
Leukaemia!

Pneumonia

Typhoid

Cellulitis

Gastroenteritis

Septic
arthritis

Urinary
Tract
Infection

Juvenile
Arthritis

Fabricated or
Induced Illness

www.PicturingMedicine.com

The causes of **Breast Disease** - *Tends to present as lumps, pain and discharge*

Cyclical Mastalgia (Fibro-cystic changes).
Benign breast lumps which fluctuate with hormonal cycles.

Duct Ectasia
Dilation of lactiferous ducts. Can mimic breast cancer. Nipple retraction, pain and discharge can occur.

Galactocele
Benign cystic tumour containing milk. Occur in the lactating breast.

Carcinoma.
Most commonly arising from the inner lining of ducts and lobules.

Fibroadenoma
"Breast Mouse" (highly mobile) Small, solid, benign, rubbery lumps.

Fat Necrosis
Damage to fat tissue leading to necrosis and a benign lump.

Duct Papilloma
Localised benign growth. (Bloody nipple discharge can occur)

Retention Cyst
Benign cystic swelling caused by blockage of the secreting mechanism.

Breast Abscess
(Don't forget TB!)

Pregnancy!
The commonest cause for changes in the breasts. Can lead to tense, heavy and uncomfortable breasts. Sometimes women describe a pricking sensation. At 2 mths the breasts are often granular or lumpy. The nipple enlarges and the areola darkens.

Don't forget
- Lipomas
- Sebaceous cysts
- Shingles
....etc!

23

The causes of **Obstructive Jaundice** (Leads to pale stools & dark urine!)

Causes can be categorised
- In the lumen
- In the wall
- Outside the lumen

TPN (Total Parenteral Nutrition)
& Pregnancy
Both can lead to cholestasis

Durgs causing Cholestasis
- Many anti-biotics
- Oral contraceptive
- Anabolic Steroids

Biliary Atresia
Presenting in neonates

Primary Biliary Cirrhosis

Pancreatic Cancer

Gallstones

Mirizzi's Syndrome
Extrinsic compression of the CBD by a gallstone impacted in Hartmann's Pouch.

Chronic Pancreatitis

Bile Duct Strictures

Ductal Carcinoma

Choledochal cysts

Schistasomiasis

Dark Urine

Pale Stools

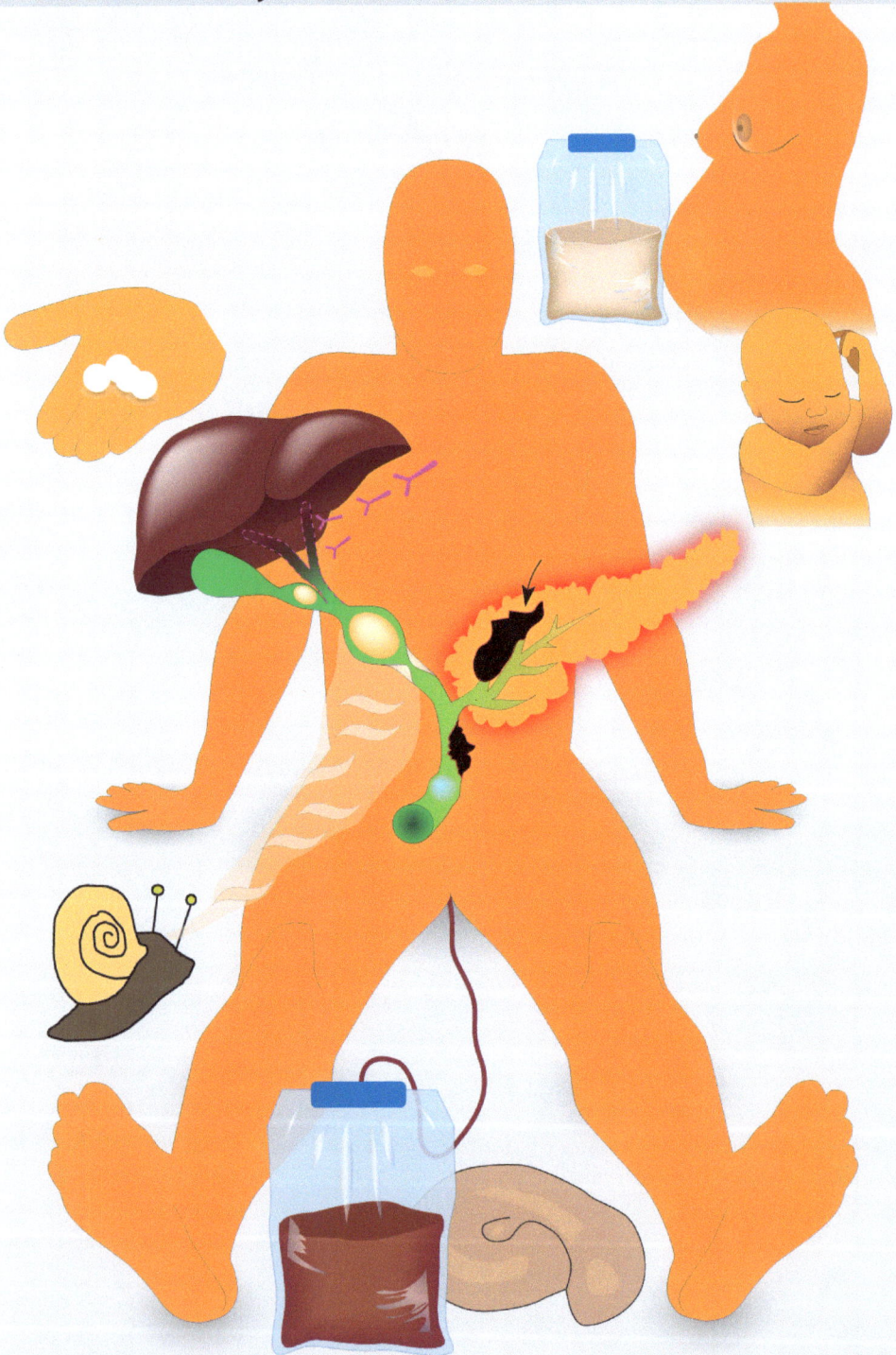

The causes of **Pancreatitis**

The mnemonic "I Get Smashed" is often found to be helpful.

The most common causes are Alcohol and Gallstones.

Idiopathic
Gallstones
Ethanol (alcohol)
Trauma
Steroids
Mumps
Autoimmune
  Diseases
Scorpion
Hypercalcaemia
  Hyperlipidaemia
  Hypothermia
ERCP
Drugs

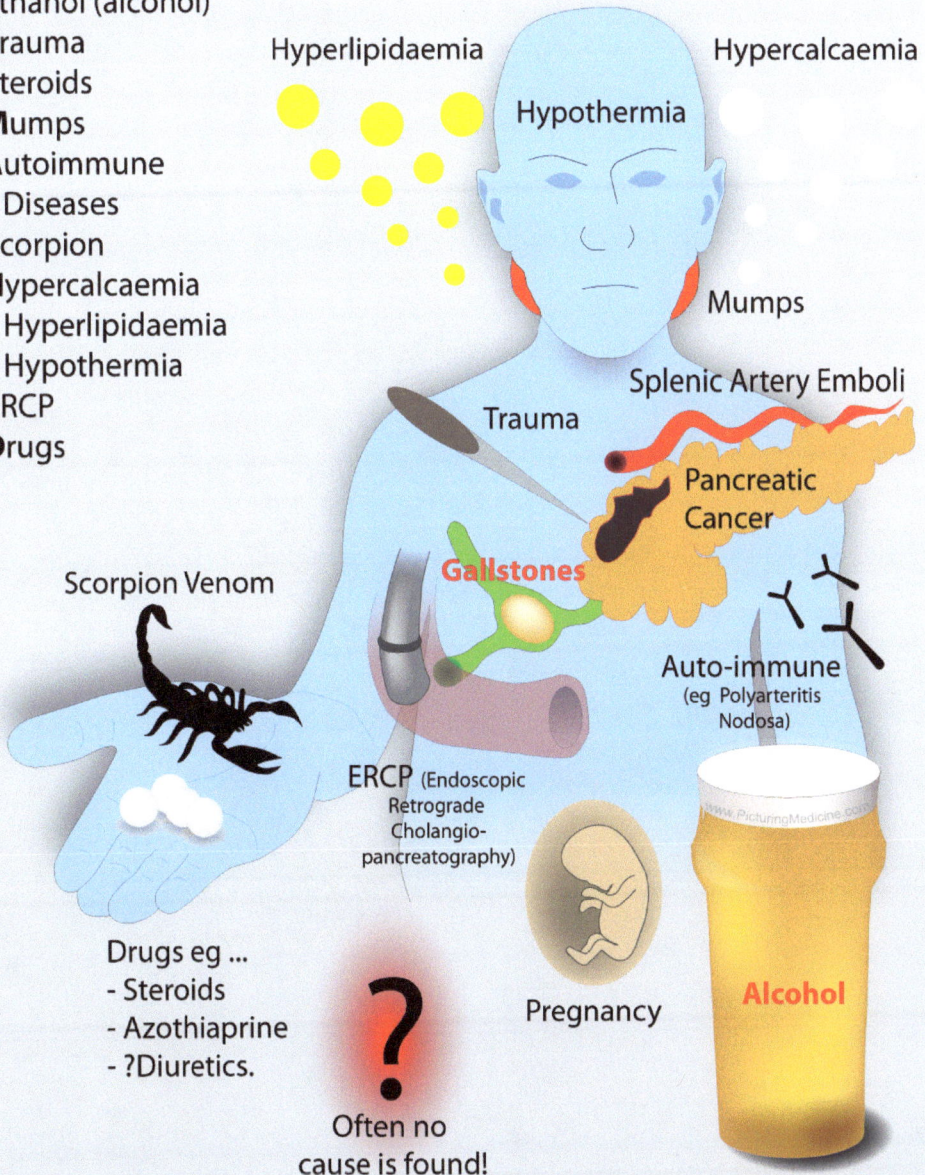

Hyperlipidaemia

Hypercalcaemia

Hypothermia

Mumps

Splenic Artery Emboli

Trauma

Pancreatic Cancer

Gallstones

Scorpion Venom

Auto-immune
(eg Polyarteritis Nodosa)

ERCP (Endoscopic Retrograde Cholangio-pancreatography)

Drugs eg ...
- Steroids
- Azothiaprine
- ?Diuretics.

?

Often no cause is found!

Pregnancy

Alcohol

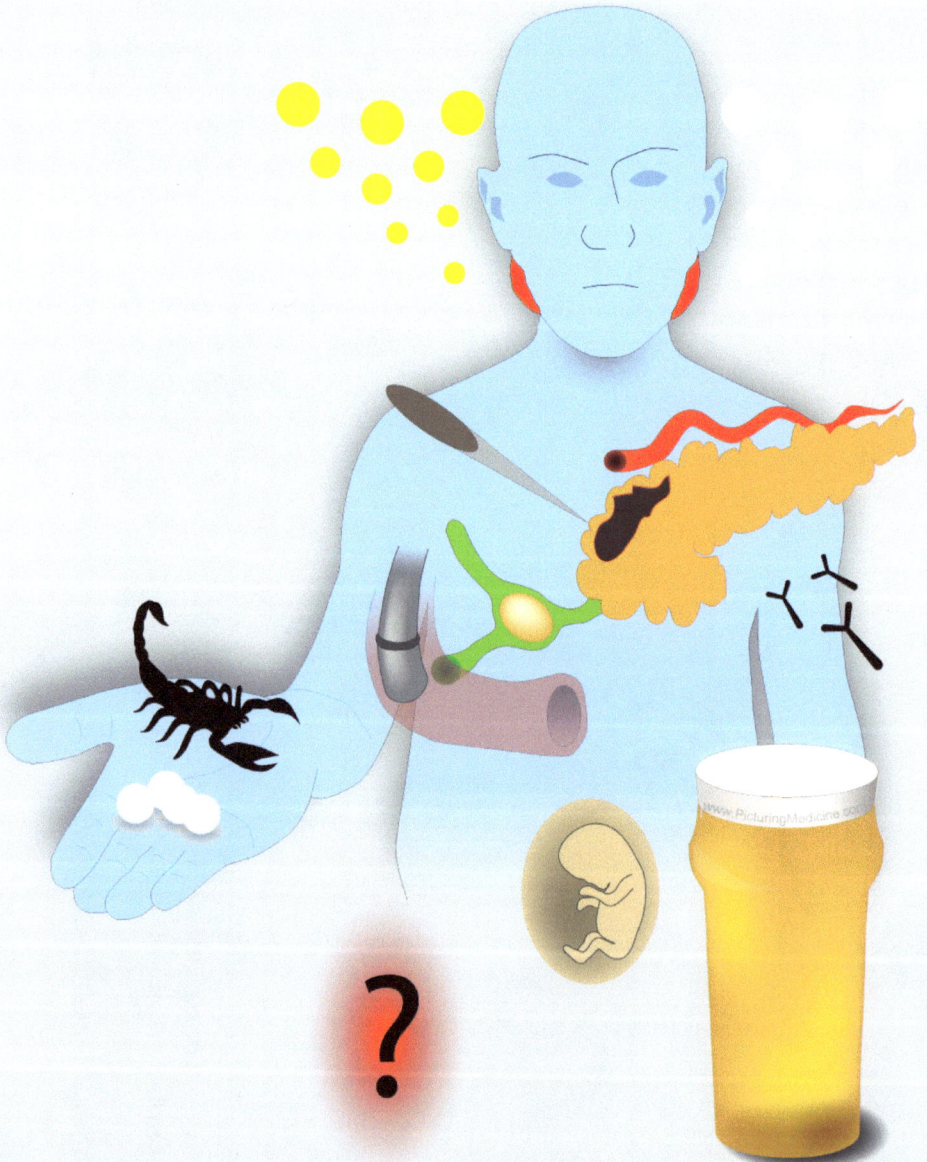

© 2010 John Kenneth Dickson

28

The causes of **Carpal Tunnel Syndrome**

Remember, carpal tunnel syndrome is most commonly idiopathic.

?

More common in ...

- Older age
- Caucasians
- Women
- High BMI & Short stature.

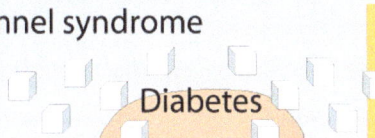

Diabetes

Acromegaly

Amyloidosis
(Beta-pleated sheets)

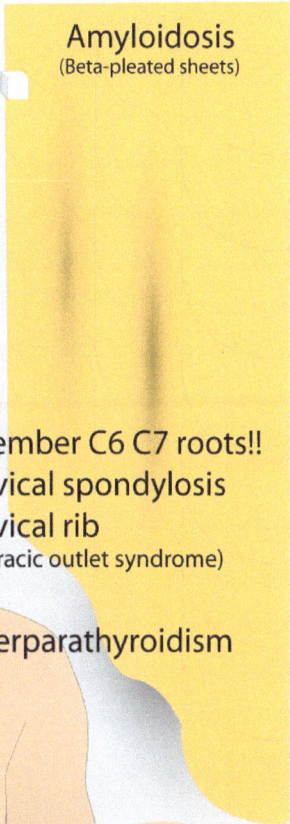

Remember C6 C7 roots!!
- Cervical spondylosis
- Cervical rib
(Thoracic outlet syndrome)

Hypothyroidism

Hyperparathyroidism

Sarcoidosis

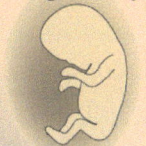

Pregnancy

Dialysis

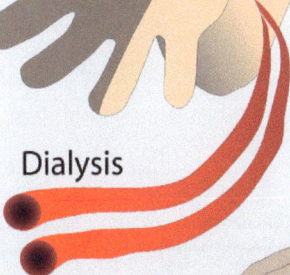

Rheumatoid Arthritis

Trauma (eg following a Colles fracture)

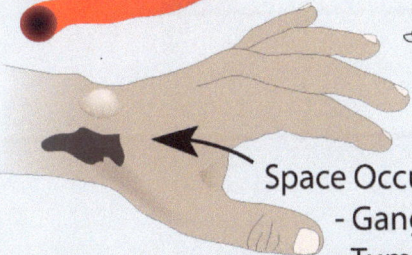

Space Occupying Lesions
- Ganglions
- Tumours

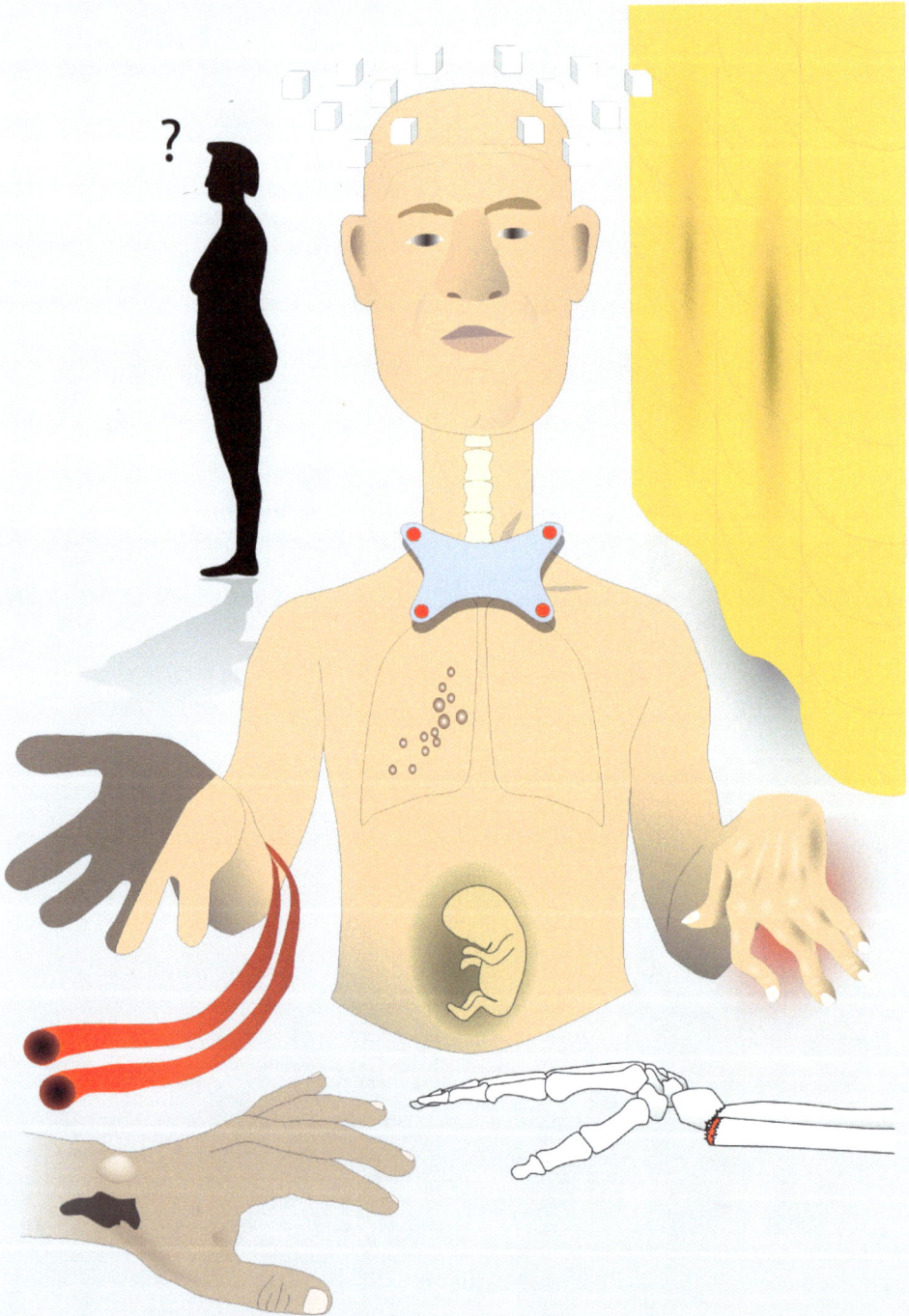

The causes of **Hand Swellings**

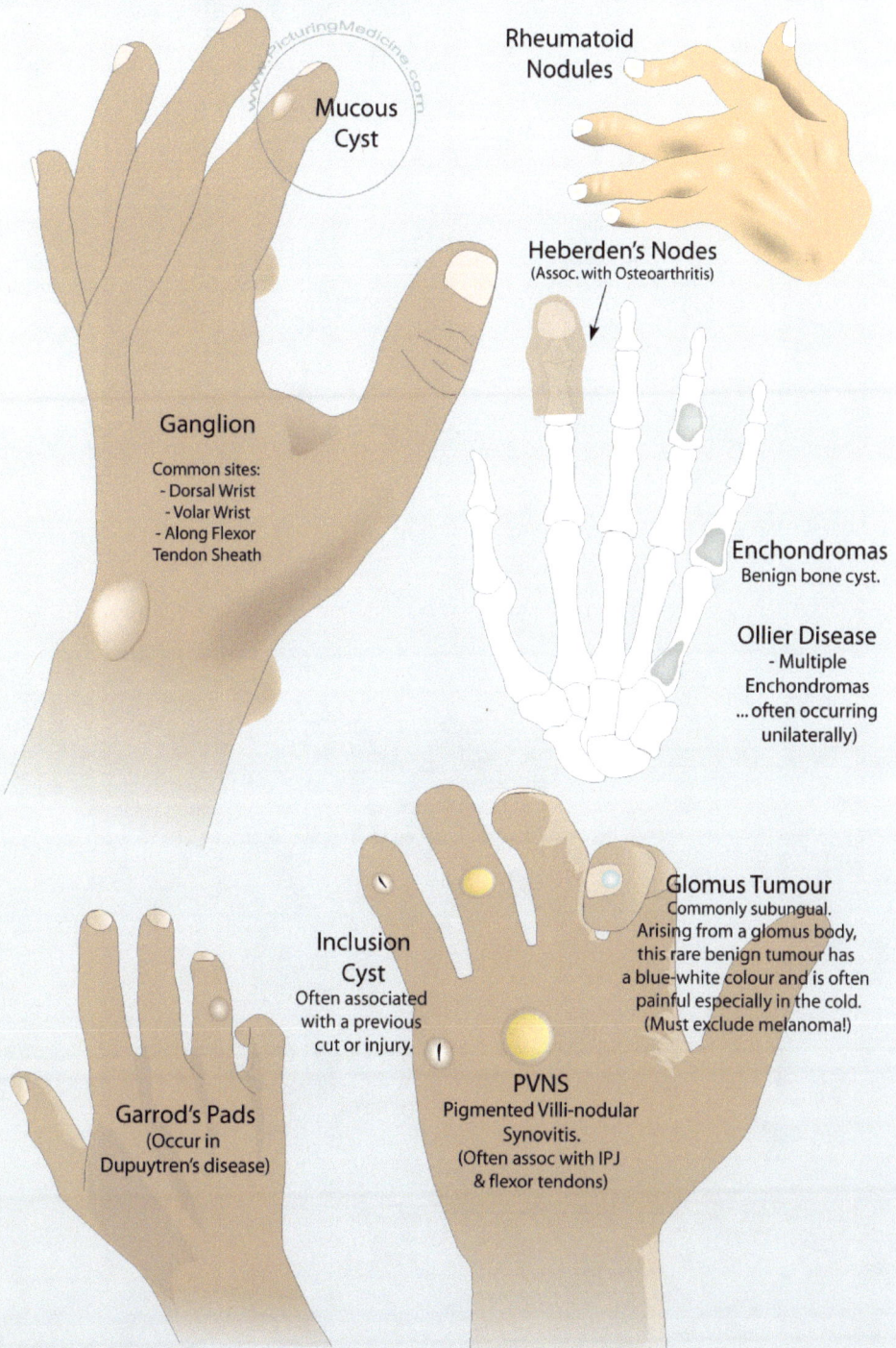

Mucous Cyst

Rheumatoid Nodules

Heberden's Nodes
(Assoc. with Osteoarthritis)

Ganglion

Common sites:
- Dorsal Wrist
- Volar Wrist
- Along Flexor Tendon Sheath

Enchondromas
Benign bone cyst.

Ollier Disease
- Multiple Enchondromas
...often occurring unilaterally)

Glomus Tumour
Commonly subungual.
Arising from a glomus body, this rare benign tumour has a blue-white colour and is often painful especially in the cold.
(Must exclude melanoma!)

Inclusion Cyst
Often associated with a previous cut or injury.

PVNS
Pigmented Villi-nodular Synovitis.
(Often assoc with IPJ & flexor tendons)

Garrod's Pads
(Occur in Dupuytren's disease)

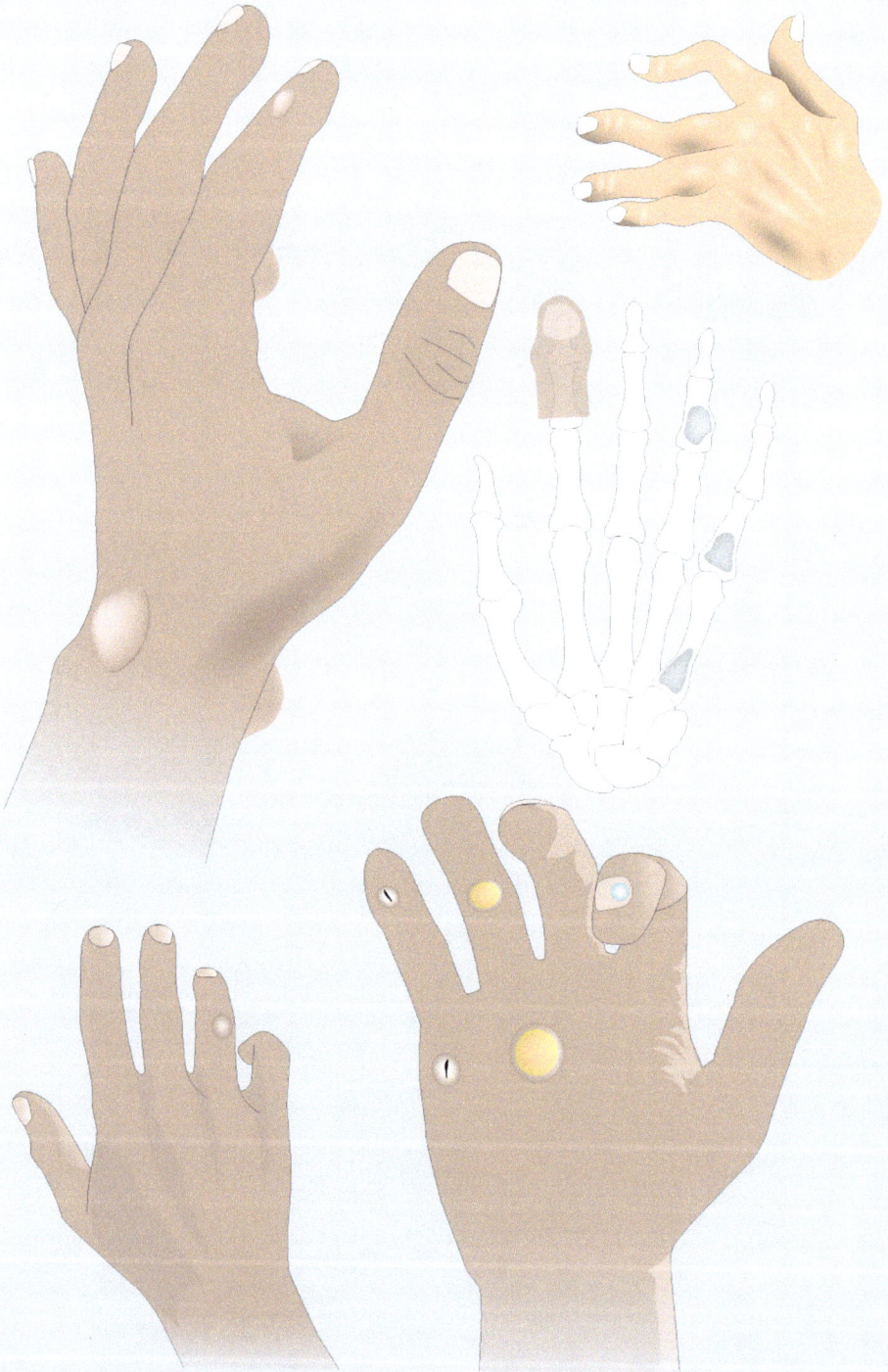

The causes of a **Swelling in the Neck**

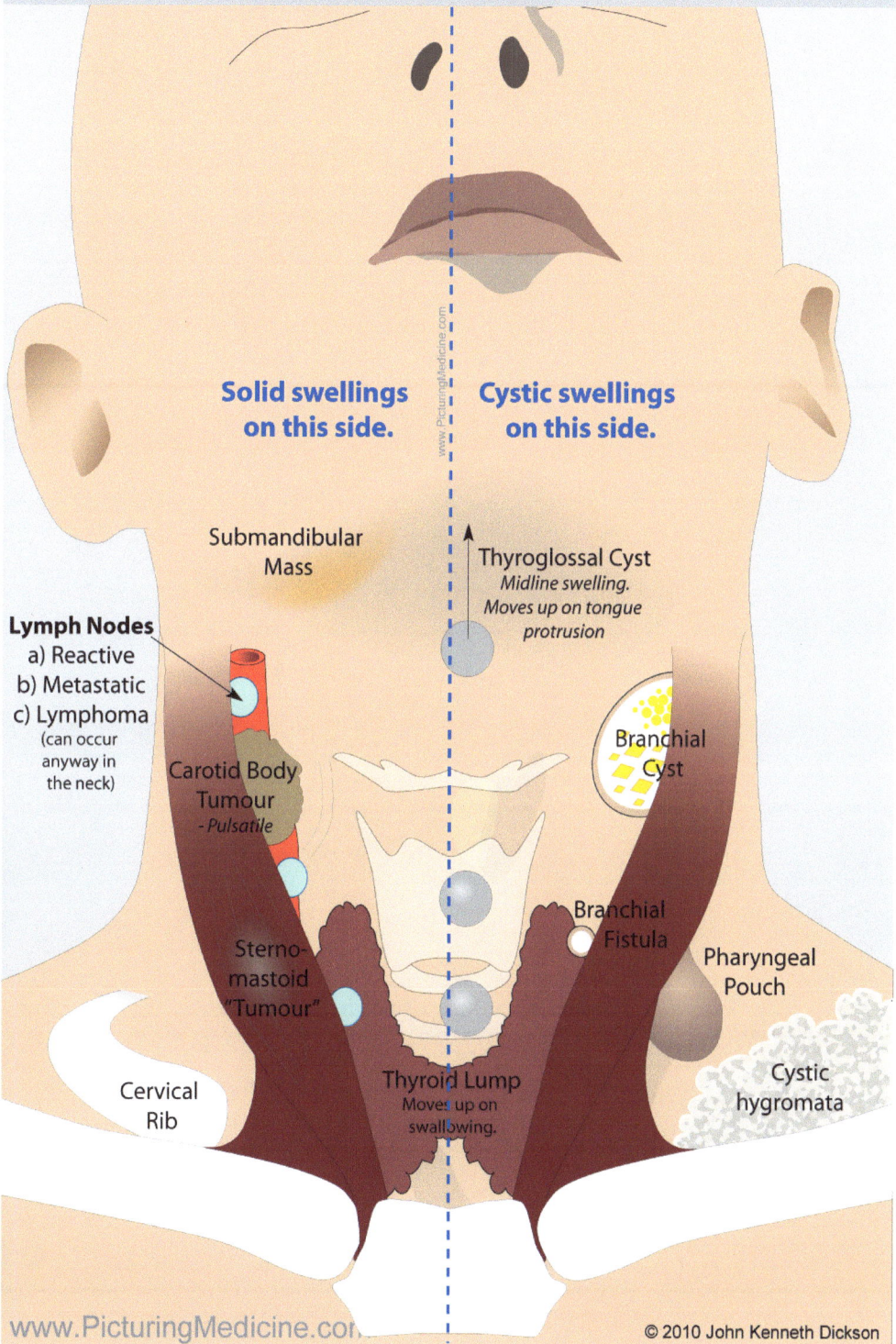

Solid swellings on this side.

Cystic swellings on this side.

www.PicturingMedicine.com

Submandibular Mass

Thyroglossal Cyst
*Midline swelling.
Moves up on tongue
protrusion*

Lymph Nodes
a) Reactive
b) Metastatic
c) Lymphoma
(can occur
anyway in
the neck)

Carotid Body
Tumour
- Pulsatile

Branchial Cyst

Sterno-
mastoid
"Tumour"

Branchial Fistula

Pharyngeal Pouch

Cervical Rib

Thyroid Lump
*Moves up on
swallowing.*

Cystic hygromata

33

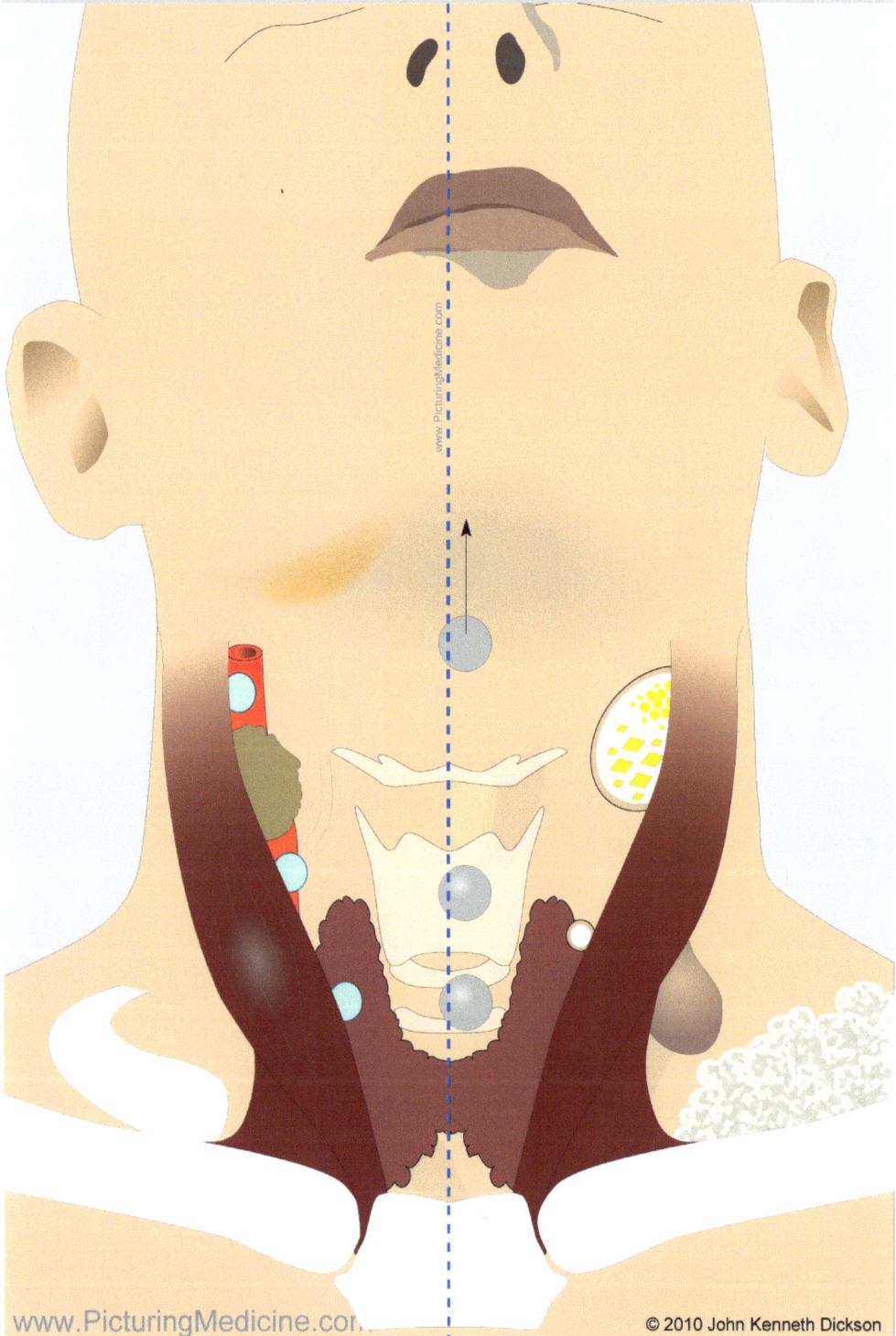

The causes of **Hypernatraemia**

Causes associated with HIGH concentration of sodium in the urine.

Causes associated with LOW concentration of sodium in the urine.

1) Simply not drinking! (Often seen in the elderly)

1) Osmotic diuresis (eg in DM)

2) Excessive Sodium Bicarbonate administration.

2) Diabetes Insipidus (Neurogenic) - Deficiency of ADH from pituitary

3) Conn's Syndrome

4) Cushing's Syndrome

3) Excessive sweating or diarrhoea in the infant

5) Diabetes Insipidus (Nephrogenic) - Renal resistence to ADH

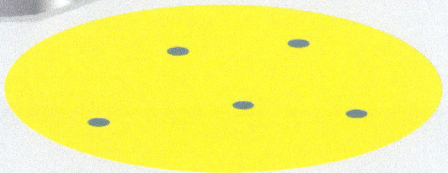

www.PicturingMedicine.com

35

The causes of **Hyponatraemia**

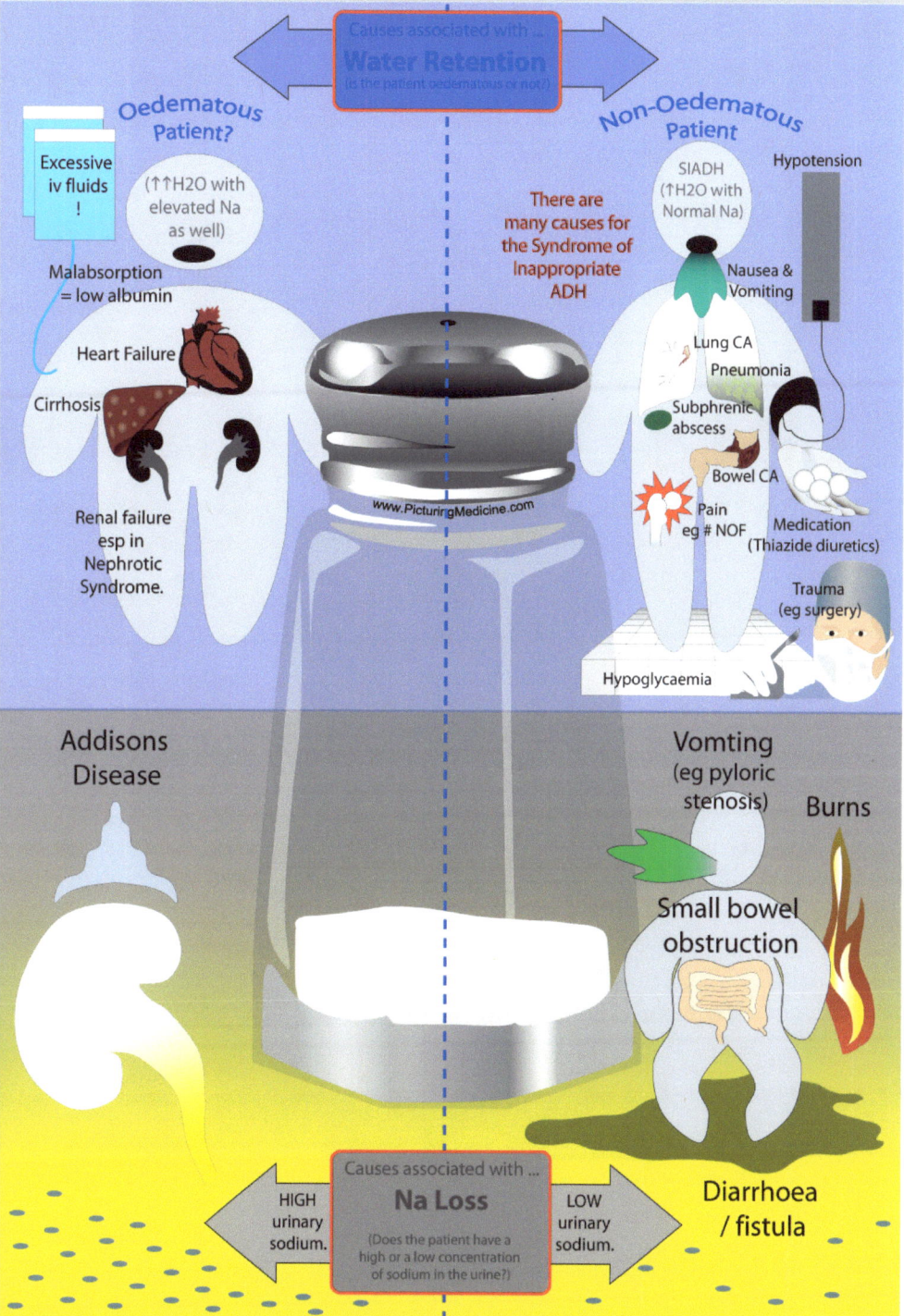

Causes associated with ...
Water Retention
(is the patient oedematous or not?)

Oedematous Patient?

Non-Oedematous Patient

Excessive iv fluids !

(↑↑H2O with elevated Na as well)

SIADH (↑H2O with Normal Na)

Hypotension

Malabsorption = low albumin

There are many causes for the Syndrome of Inappropriate ADH

Nausea & Vomiting

Heart Failure

Lung CA

Pneumonia

Cirrhosis

Subphrenic abscess

Bowel CA

Renal failure esp in Nephrotic Syndrome.

Pain eg # NOF

Medication (Thiazide diuretics)

www.PicturingMedicine.com

Trauma (eg surgery)

Hypoglycaemia

Addisons Disease

Vomting (eg pyloric stenosis)

Burns

Small bowel obstruction

Causes associated with ...
Na Loss
(Does the patient have a high or a low concentration of sodium in the urine?)

HIGH urinary sodium.

LOW urinary sodium.

Diarrhoea / fistula

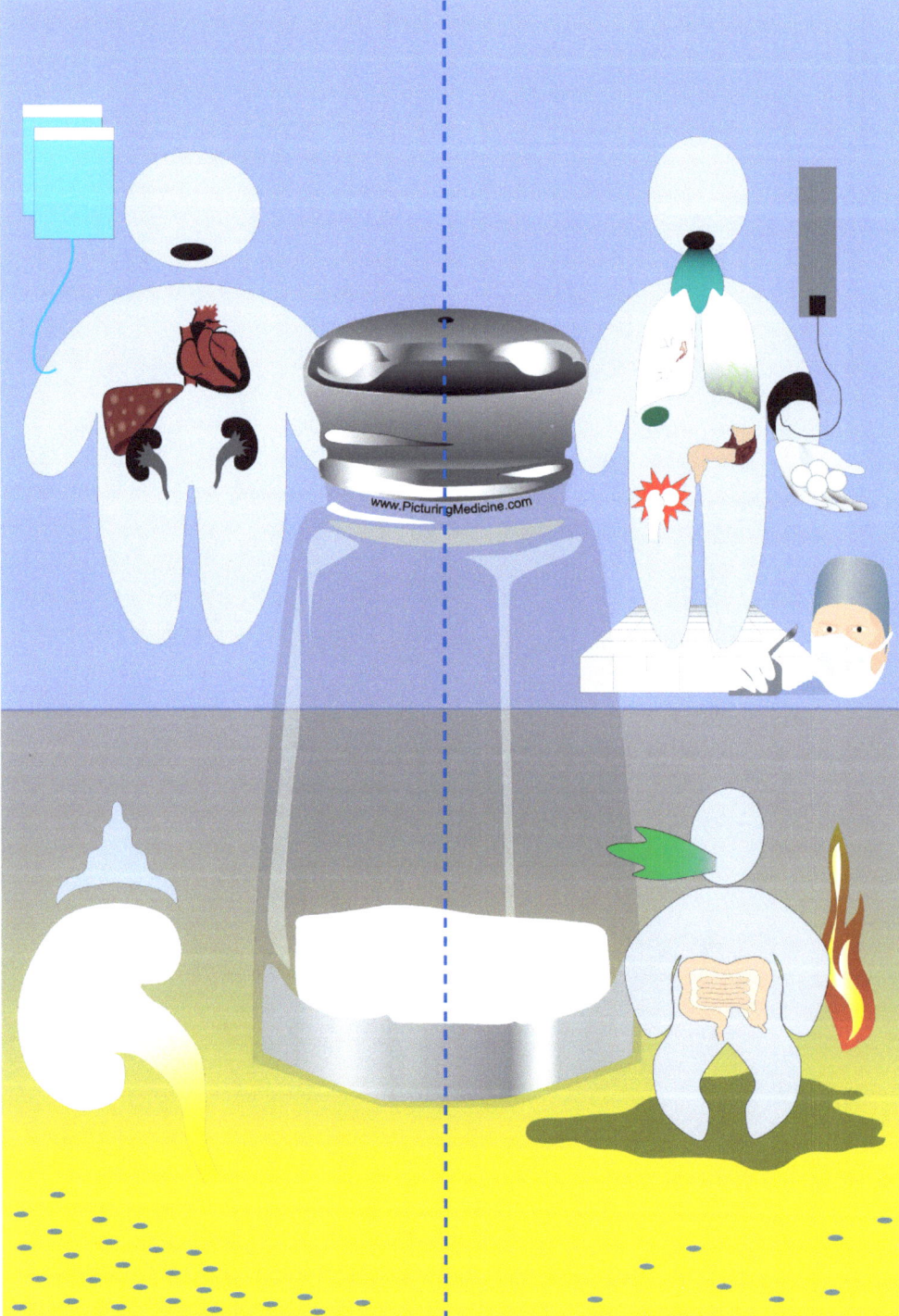

www.PicturingMedicine.com

The causes of **Hypercalcaemia**

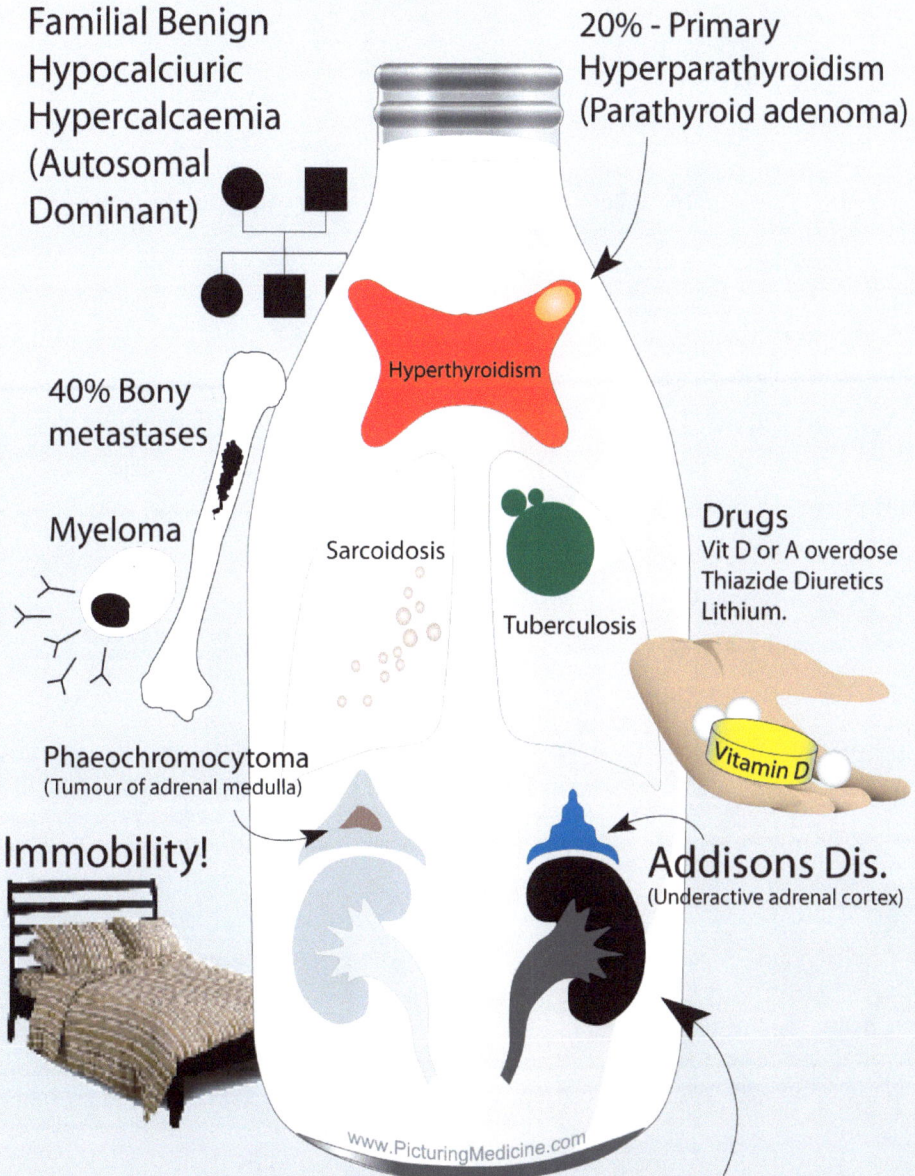

Familial Benign Hypocalciuric Hypercalcaemia (Autosomal Dominant)

20% - Primary Hyperparathyroidism (Parathyroid adenoma)

Hyperthyroidism

40% Bony metastases

Myeloma

Sarcoidosis

Tuberculosis

Drugs
Vit D or A overdose
Thiazide Diuretics
Lithium.

Vitamin D

Phaeochromocytoma
(Tumour of adrenal medulla)

Immobility!

Addisons Dis.
(Underactive adrenal cortex)

www.PicturingMedicine.com

Tertiary Hyperparathyroidism :-
Renal failure causes low Ca leading to secondary hyperparathyroidism
Sometimes, when the renal failure has resolved or been treated the parathyroid
glands continue to secrete inappropriately high levels of parathyroid hormone
leading to tertiary hyperparathyroidism with associated elevated Ca levels.

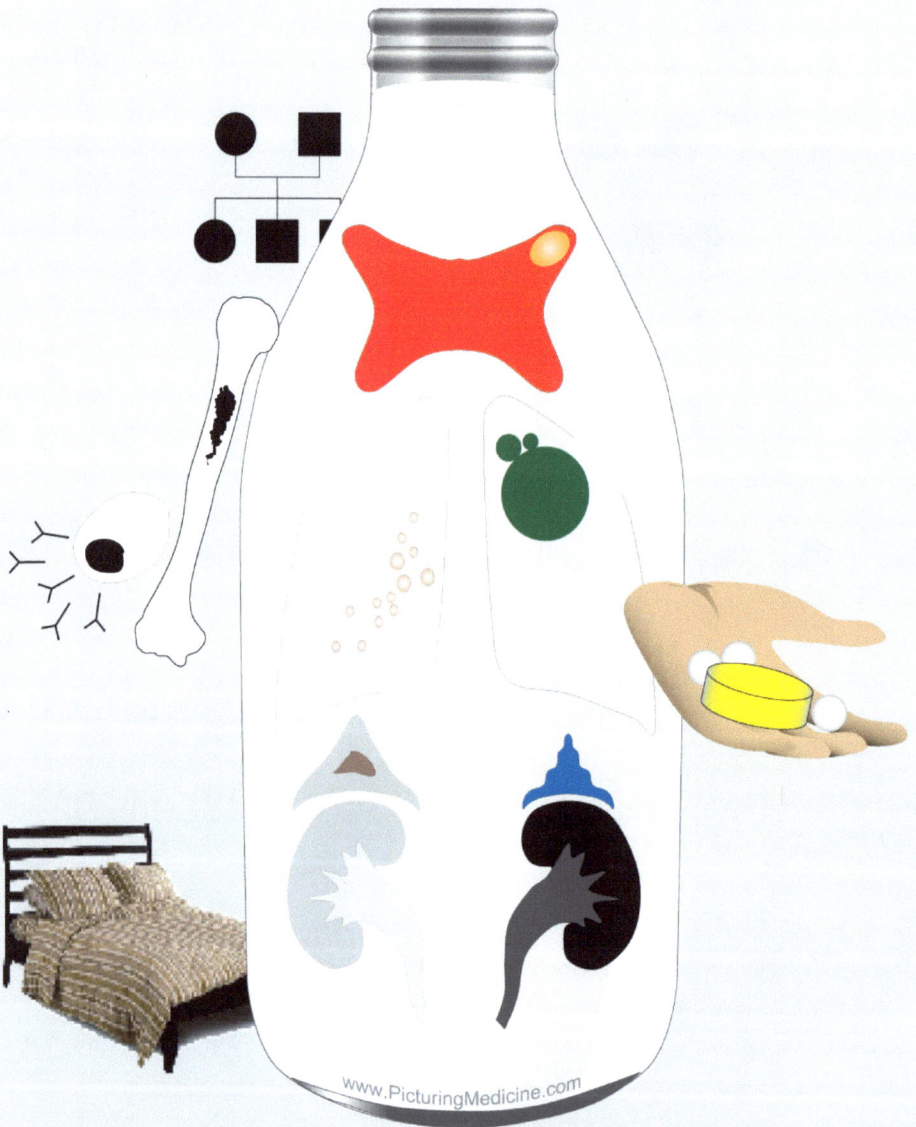

www.PicturingMedicine.com

The causes of **Hypocalcaemia**

Acute Causes

Chronic Causes

Vitamin D deficiency
(dietary lack / lack of sunlight)

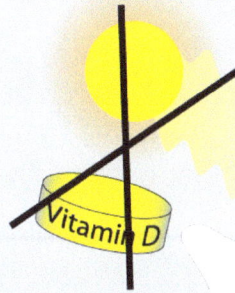

Vitamin D

Hyperventilation
*"Blowing off CO2 leads to ...
Respiratory Alkalosis*

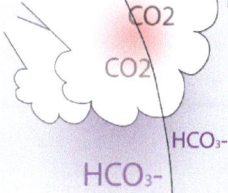

CO_2

CO_2

HCO_3^-

HCO_3^-

Thyroid

Hypoparathyroidism
- Inherited / sporadic
- **Iatrogenic** (surgical removal)

Acute
Pancreatitis

Chronic Renal Disease

Malabsorption

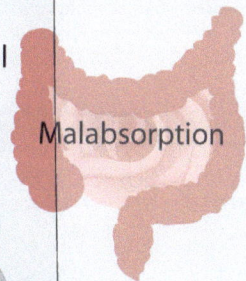

HCO_3^-

HCO_3^-

Salicylate

HCO_3^-

Metabolic
Alkalosis
eg Salicylate
poisoning or
Alkali infusion
to treat
acidosis!

Low Mg
- *Causes low calcium due
to the fact that it reduces
PTH release.*

Mg

Mg

Mg

Loop Diuretics

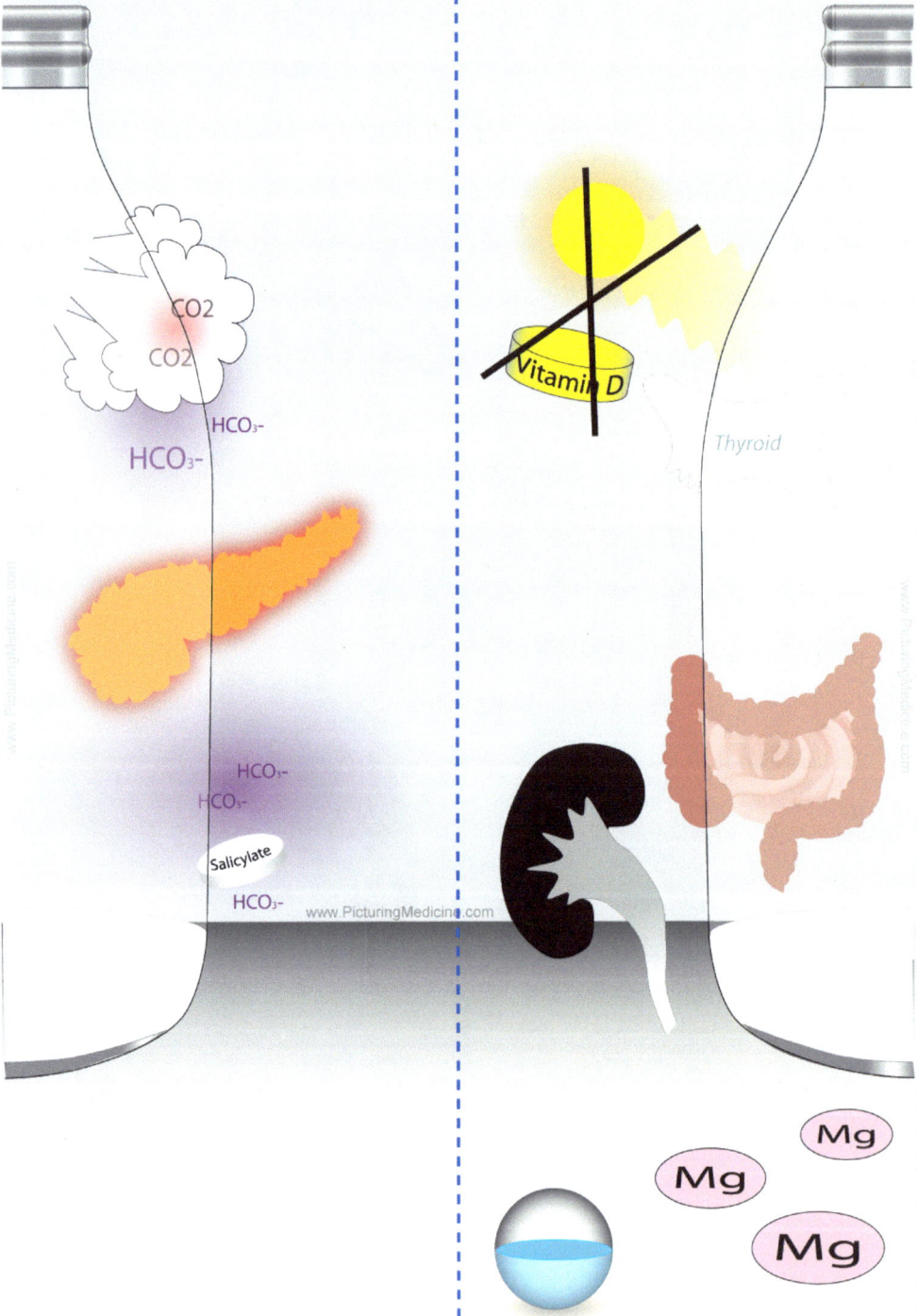

CO2

CO2

HCO₃-

HCO₃-

Vitamin D

Thyroid

HCO₃-

HCO₃-

Salicylate

HCO₃-

Mg

Mg

Mg

The causes of **Hyperkalaemia**

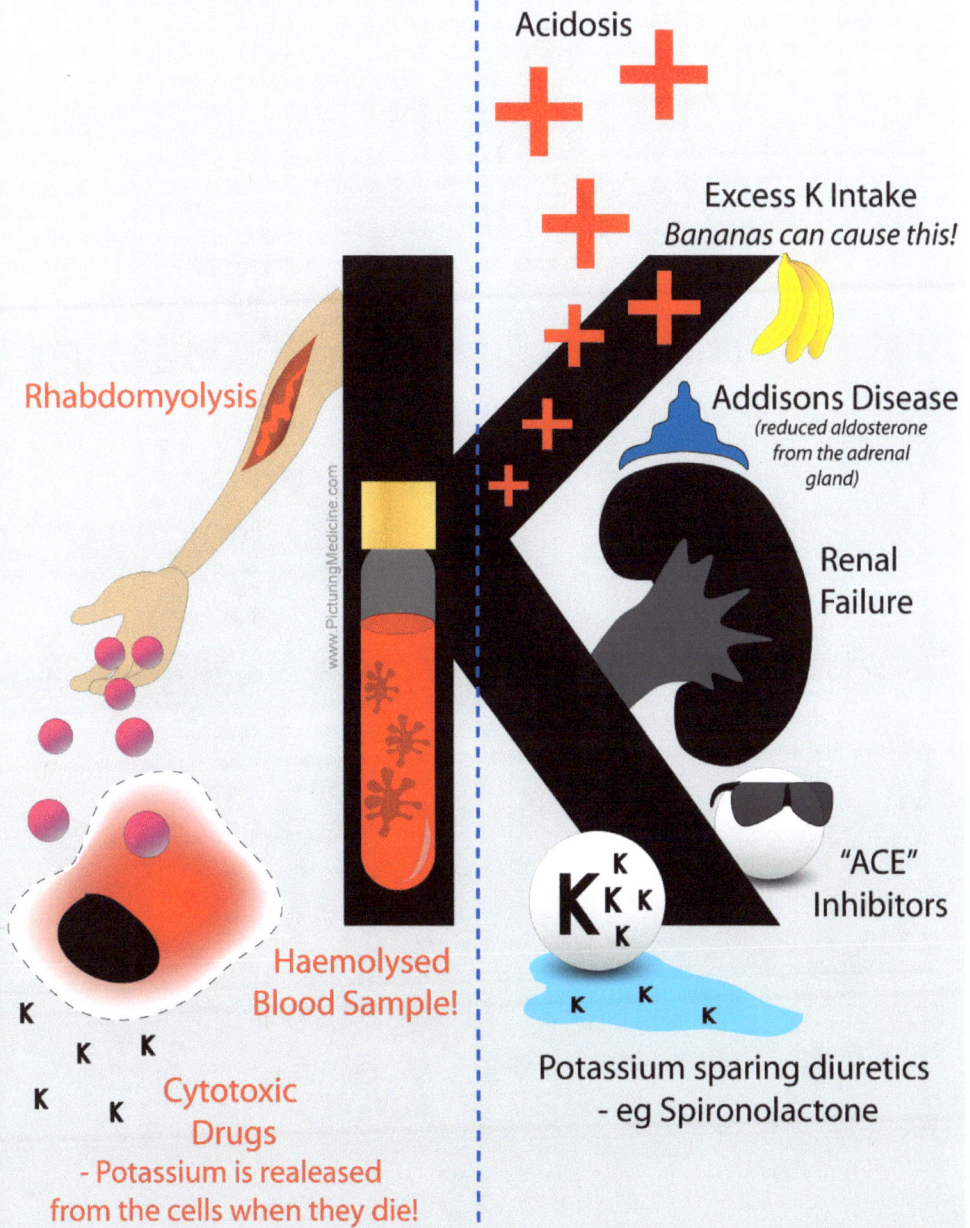

Causes associated with HAEMOLYSIS

Other causes

Rhabdomyolysis

Haemolysed Blood Sample!

K K K
K K
Cytotoxic Drugs
- Potassium is realeased from the cells when they die!

Acidosis

Excess K Intake
Bananas can cause this!

Addisons Disease
(reduced aldosterone from the adrenal gland)

Renal Failure

"ACE" Inhibitors

K K K K

K K K

Potassium sparing diuretics - eg Spironolactone

43

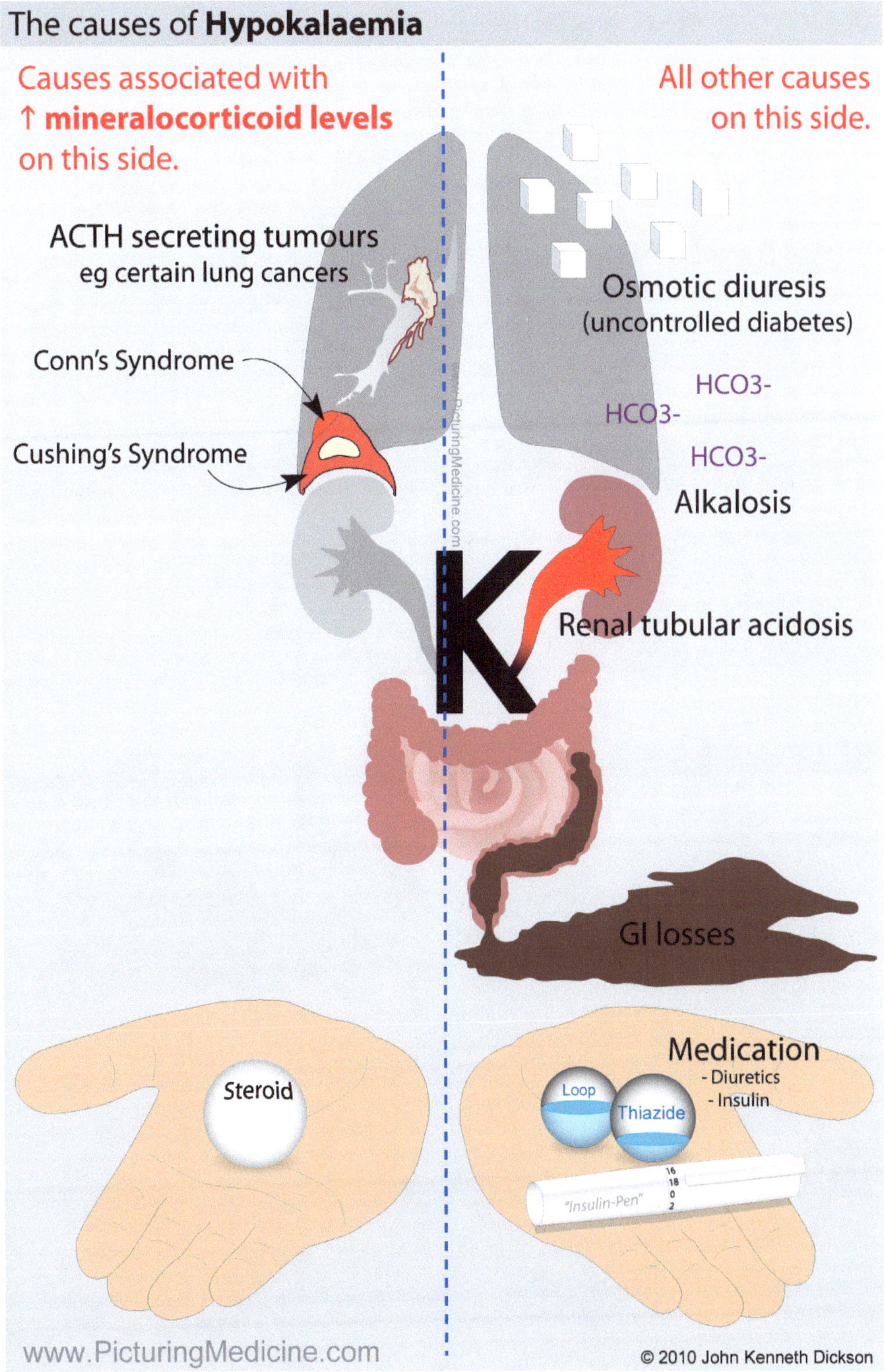

The causes of **Hypokalaemia**

Causes associated with
↑ **mineralocorticoid levels**
on this side.

All other causes
on this side.

ACTH secreting tumours
eg certain lung cancers

Osmotic diuresis
(uncontrolled diabetes)

Conn's Syndrome

HCO3-
HCO3-

Cushing's Syndrome

HCO3-

Alkalosis

K

Renal tubular acidosis

GI losses

Medication
- Diuretics
- Insulin

Steroid

Loop
Thiazide

16
18
0
2

"Insulin-Pen"

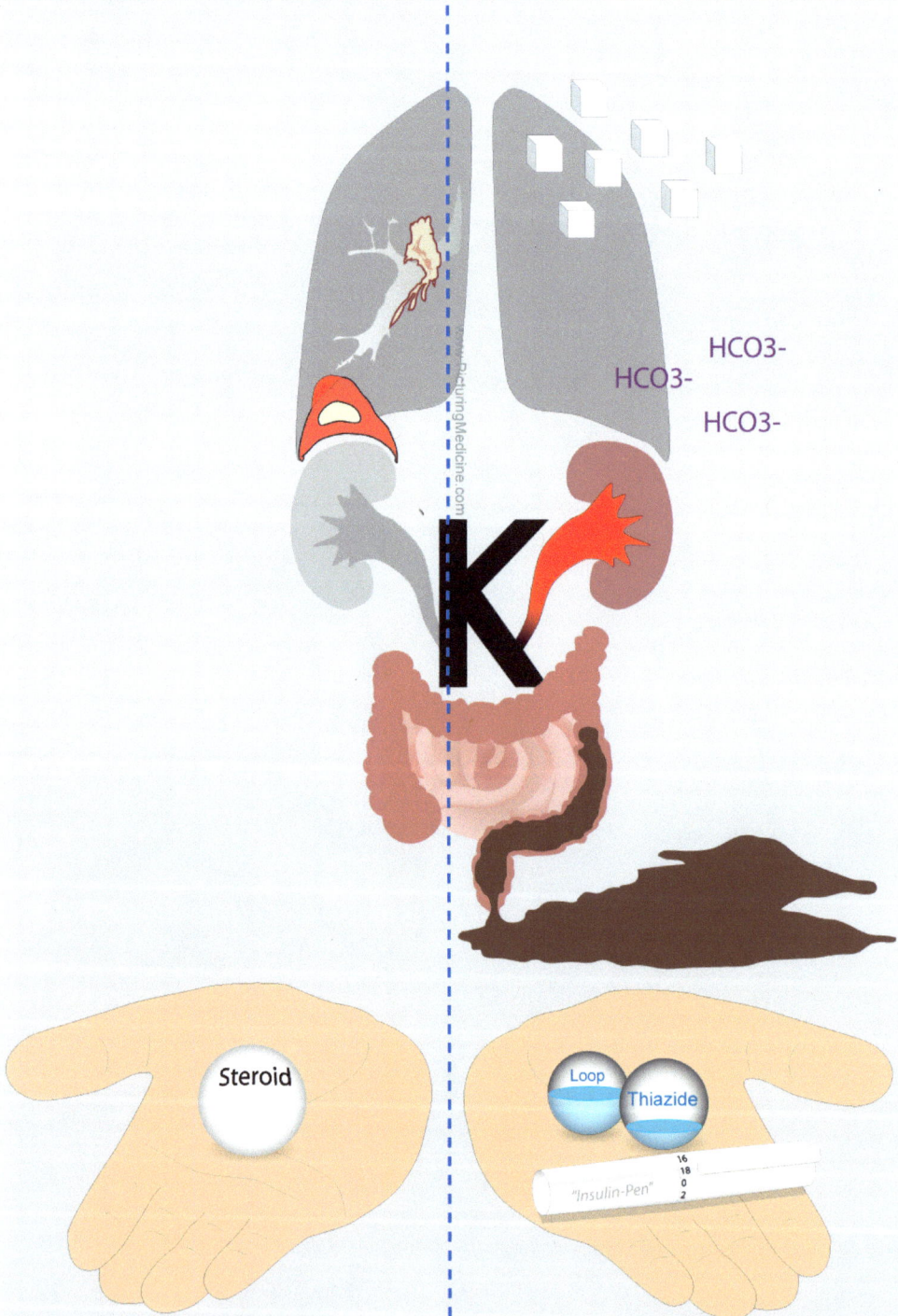

HCO3-

HCO3-

HCO3-

K

Steroid

Loop

Thiazide

16
18
0
2

"Insulin-Pen"

The causes of **Anaemia** - *Grouped into Microcytic, Normocytic & Macrocytic*

Microcytic

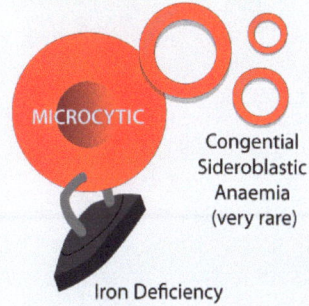

Thalassaemia
Common in the
mediterranean

MICROCYTIC

Congential
Sideroblastic
Anaemia
(very rare)

Iron Deficiency

Normocytic

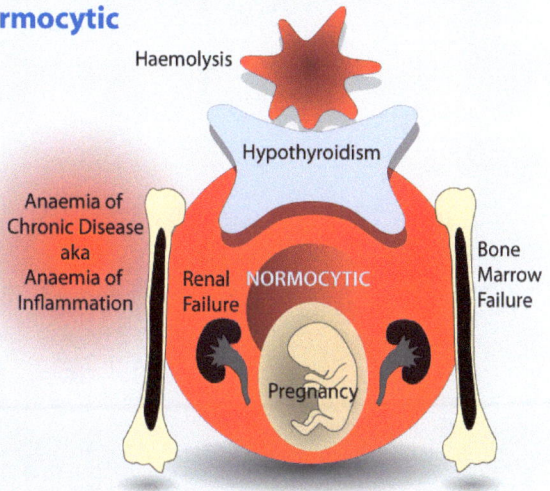

Haemolysis

Hypothyroidism

Anaemia of
Chronic Disease
aka
Anaemia of
Inflammation

Renal
Failure

NORMOCYTIC

Bone
Marrow
Failure

Pregnancy

Macrocytic

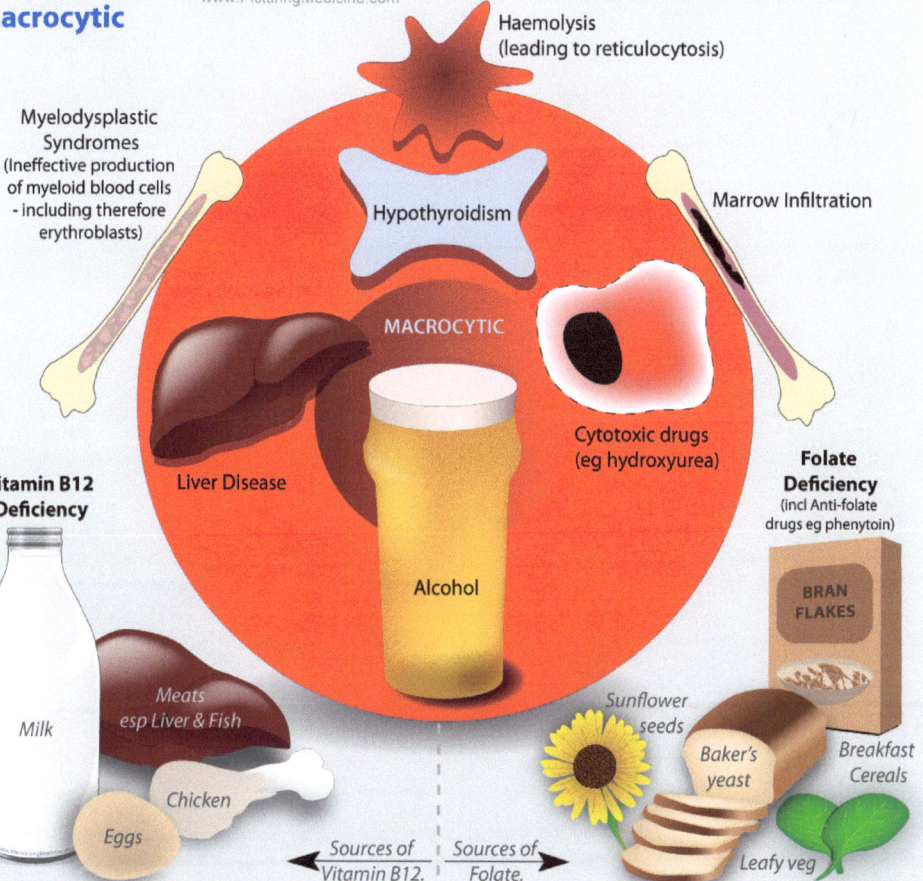

Haemolysis
(leading to reticulocytosis)

Myelodysplastic
Syndromes
(Ineffective production
of myeloid blood cells
- including therefore
erythroblasts)

Hypothyroidism

Marrow Infiltration

MACROCYTIC

**Vitamin B12
Deficiency**

Liver Disease

Cytotoxic drugs
(eg hydroxyurea)

**Folate
Deficiency**
(incl Anti-folate
drugs eg phenytoin)

BRAN
FLAKES

Alcohol

Milk

*Meats
esp Liver & Fish*

Chicken

Eggs

*Sunflower
seeds*

*Baker's
yeast*

*Breakfast
Cereals*

Leafy veg

← *Sources of
Vitamin B12.*

*Sources of
Folate.* →

47

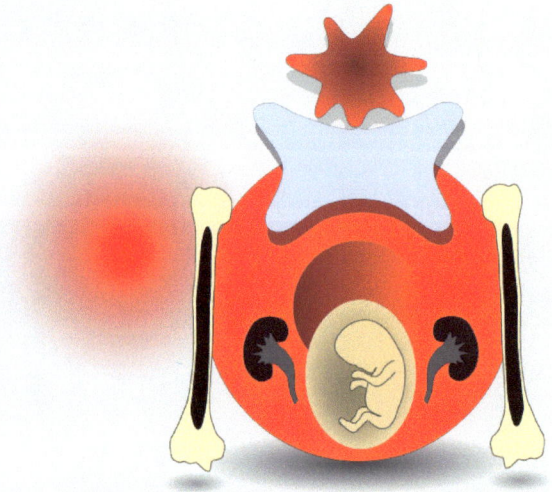

www.PicturingMedicine.com

BRAN FLAKES

The causes of **Ascites**

- ■ - High Protein Levels in Ascites
- ■ - Low Protein Levels in Ascites
  Generally associated with Portal Hypertnesion
  Causes can be pre-, intra- or post-hepatic.

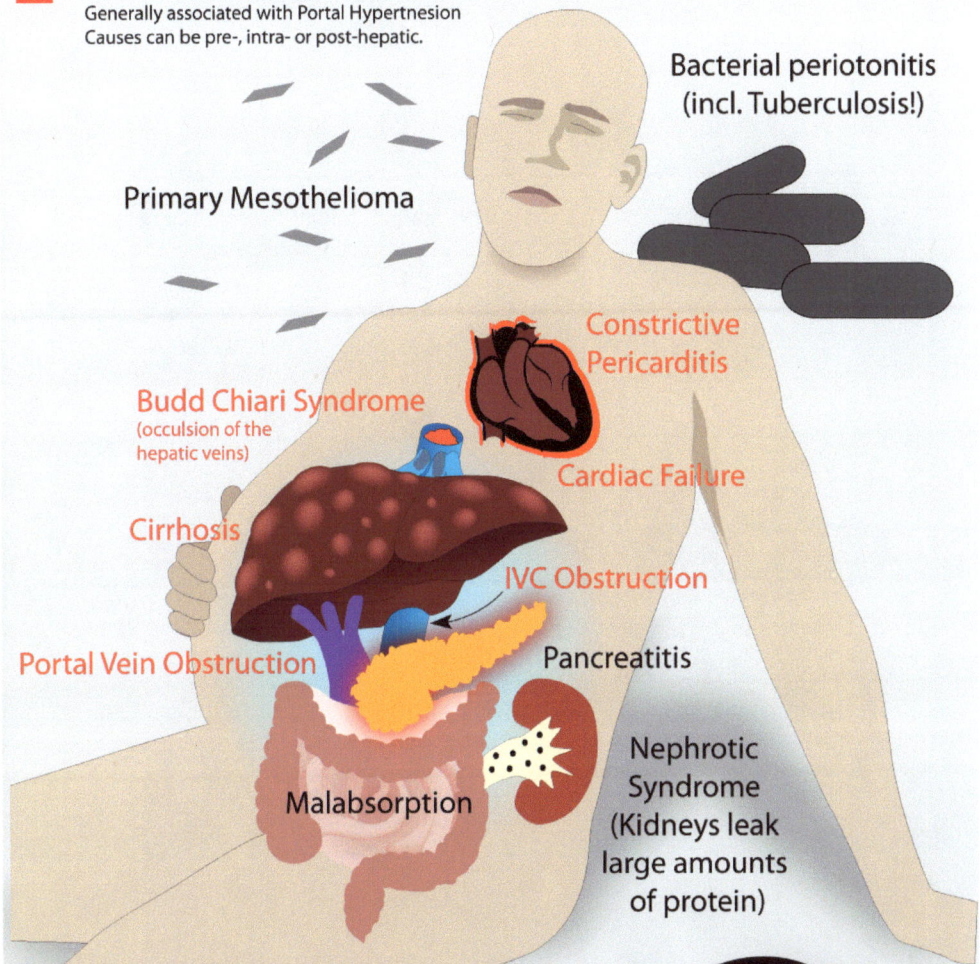

Bacterial periotonitis (incl. Tuberculosis!)

Primary Mesothelioma

Constrictive Pericarditis

Budd Chiari Syndrome (occulsion of the hepatic veins)

Cardiac Failure

Cirrhosis

IVC Obstruction

Portal Vein Obstruction

Pancreatitis

Malabsorption

Nephrotic Syndrome (Kidneys leak large amounts of protein)

CANCER: -
Carcinomatosis (widespread cancer)
Abdominal / Pelvic Cancer (Primary or Secondary)
Hepatic Tumours
Pseudomyxoma peritonei
(Uncommon tumour - characteristic mucin
spread throughout the abdomen)

49

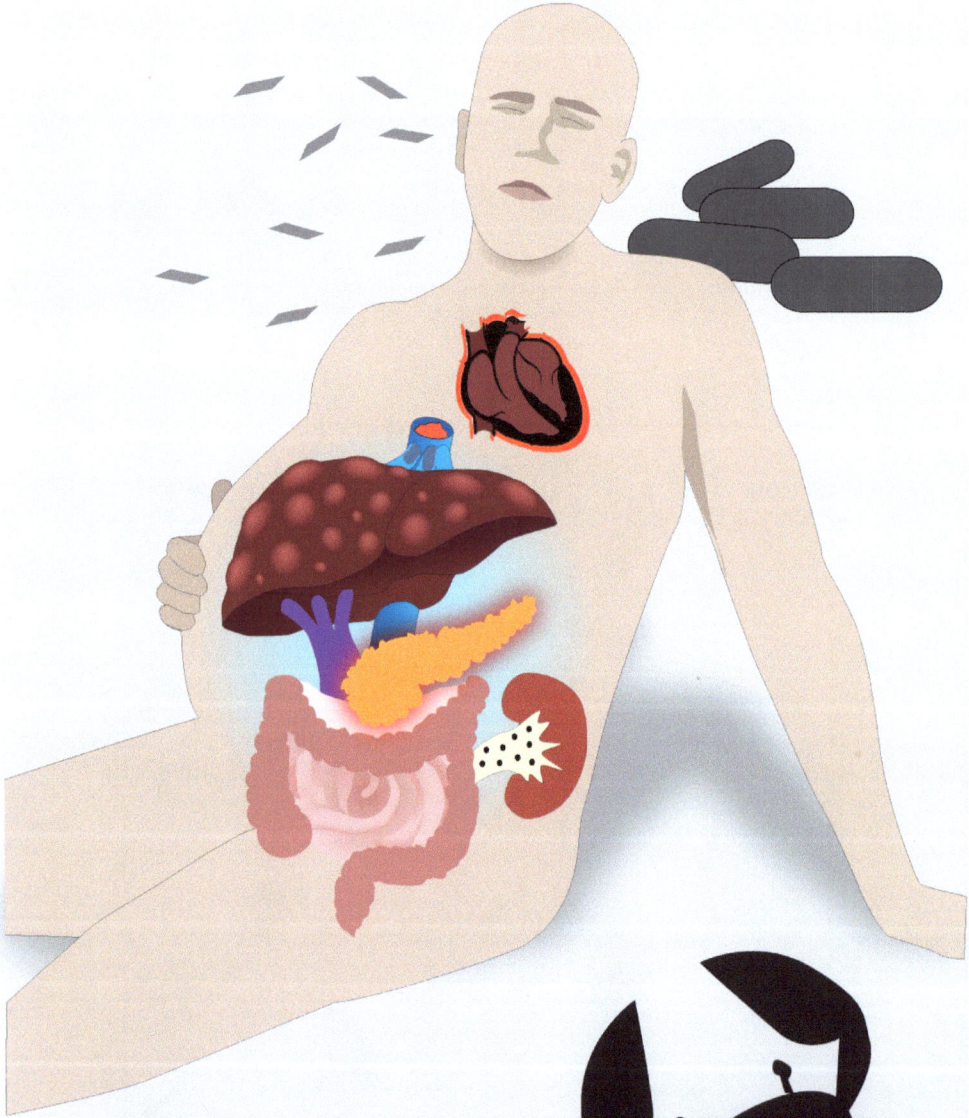

The causes of **Hypertension**

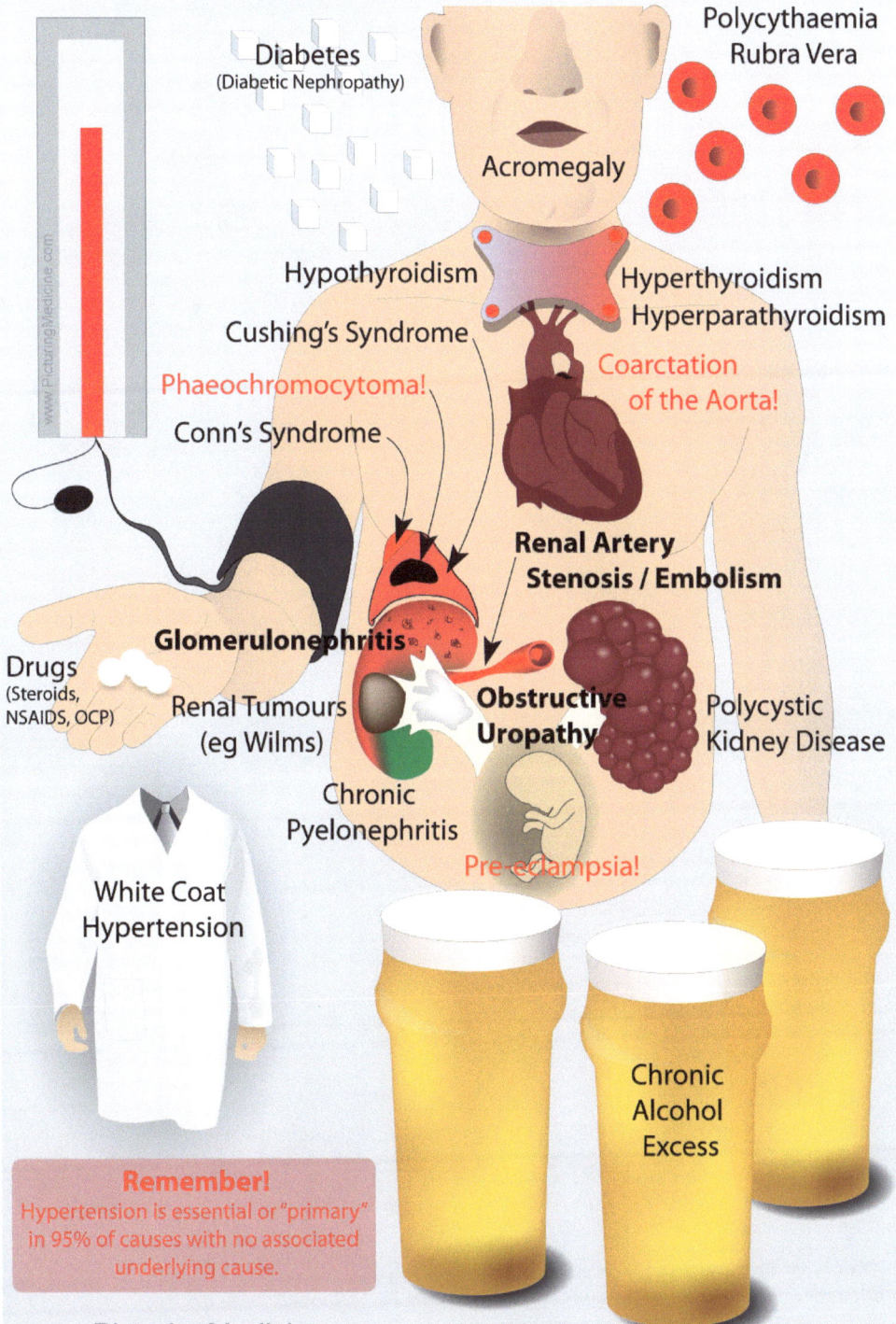

Diabetes
(Diabetic Nephropathy)

Polycythaemia
Rubra Vera

Acromegaly

Hypothyroidism

Hyperthyroidism
Hyperparathyroidism

Cushing's Syndrome

Coarctation
of the Aorta!

Phaeochromocytoma!

Conn's Syndrome

**Renal Artery
Stenosis / Embolism**

Glomerulonephritis

Drugs
(Steroids,
NSAIDS, OCP)

Renal Tumours
(eg Wilms)

**Obstructive
Uropathy**

Polycystic
Kidney Disease

Chronic
Pyelonephritis

Pre-eclampsia!

White Coat
Hypertension

Chronic
Alcohol
Excess

Remember!
Hypertension is essential or "primary"
in 95% of causes with no associated
underlying cause.

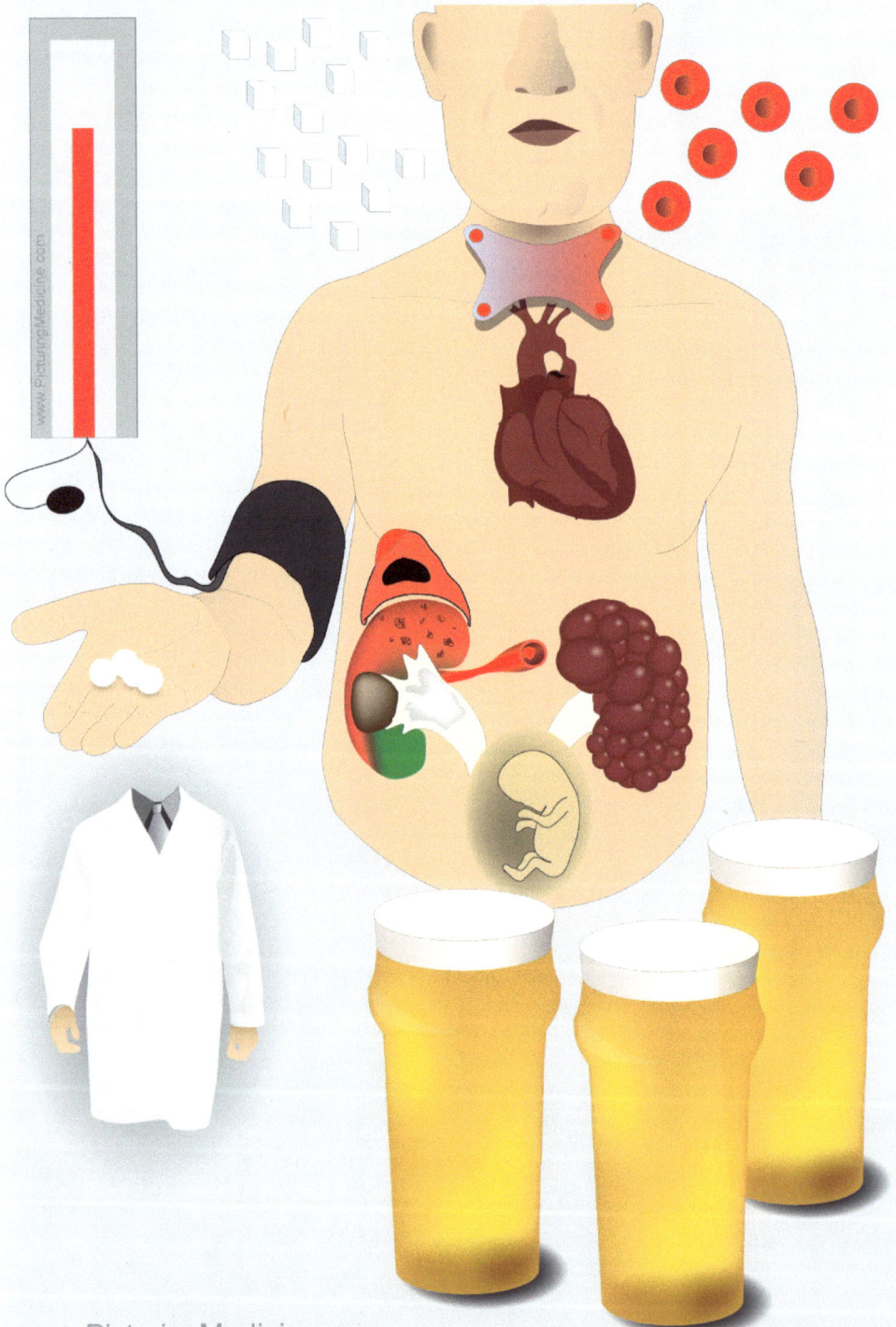

52

The causes of **Chronic Kidney Disease**

Chronic Glomerulonephritis 16%

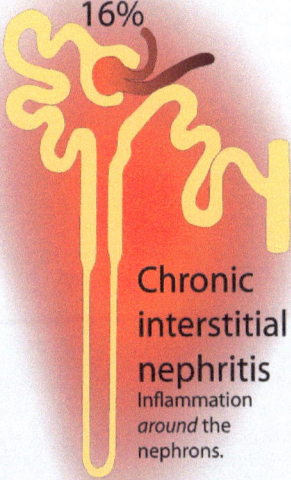

Idiopathic! 23%

?

Amyloidosis
β-pleated sheets

Chronic interstitial nephritis
Inflammation *around* the nephrons.

Diabetes 12%

Hypertension 7%

Nephrocalcinosis

Stones

Adult polycystic kidney dis 4%

Renovascular disease 9%

Tumour

Lysosome

Chronic pyelonephritis 13%
(infection)

Fabry's Dis
Lysosomal storage dis

Gout

Drugs / Toxins

Vasculitis

Port

Myeloma

Alport syndrome
deafness + renal failure

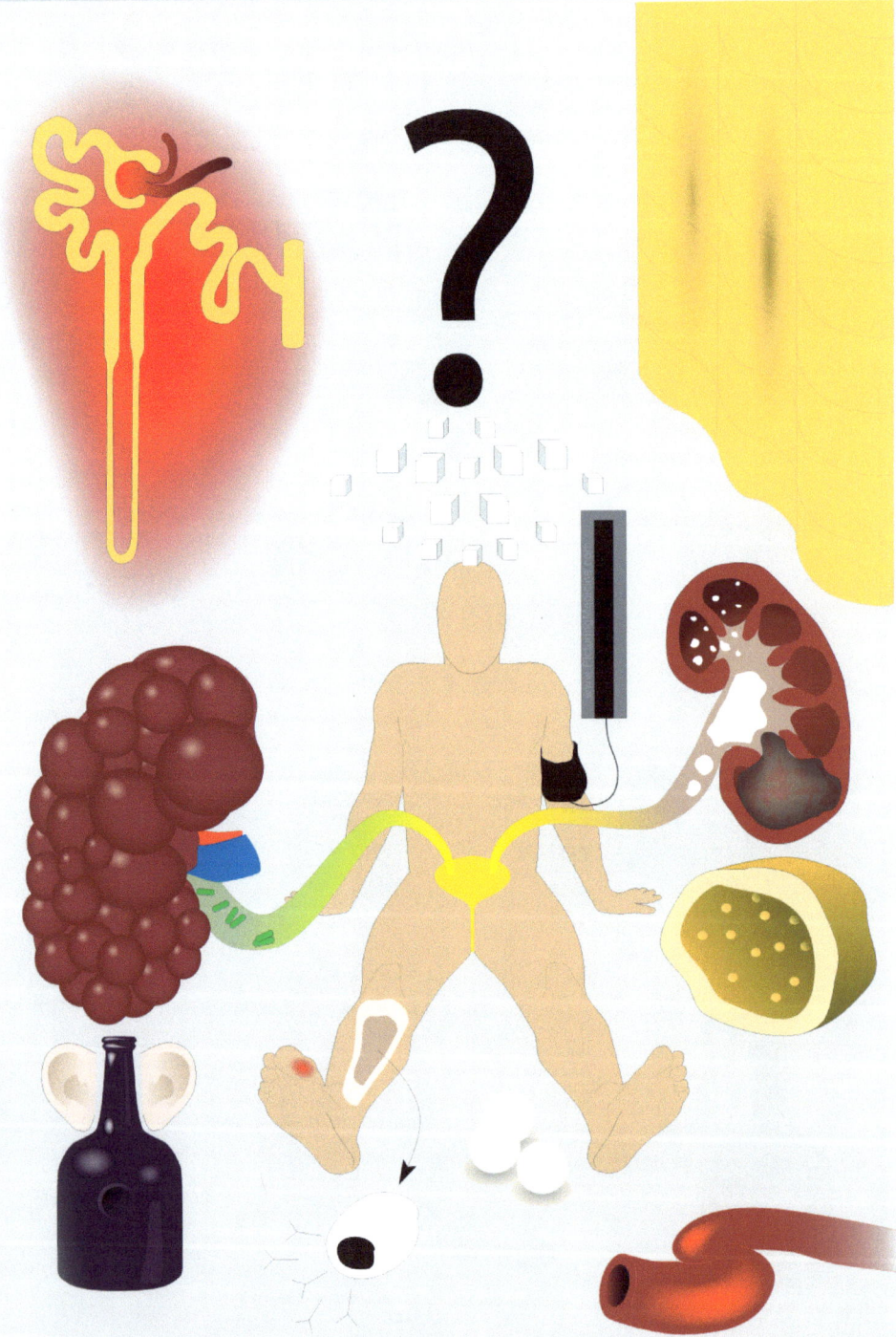

The causes of **Splenomegaly**

Haematological Malignancy

- Polycythaemia Rubra Vera
- CLL
- Myelofibrosis
- CML
- Multiple Myeloma

www.MemorizingMedicine.com

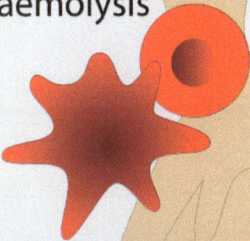

Storage diseases!
- Gaucher's dis
- Neimann-Pick dis
- Tay-Sach's dis

Amyloidosis
β-pleated sheets

Viral...
Epstein Barr Virus
Glandular Fever

Parasites...
Trypanosomiasis
Spread by the Reduviid Bug in Latin America

Schistosomiasis

Malaria & Kala-azar
Spread by the mosquito

Lymphoma

Felty syndrome
Rheumatoid Arthritis
+ Splenomegaly
+ Low WBC count.

Sarcoidosis

Infective endocarditis

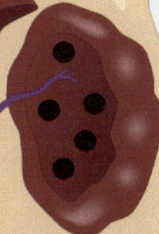

Bacteria...
Tuberculosis

Typhoid
Flagellated Gram neg bacillus

Portal Hypertension

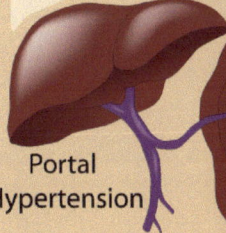

Metastases

Brucellosis
Common in Kuwait & Saudi Arabia

KUWAIT

Haemolysis

© 2010 John Kenneth Dickson

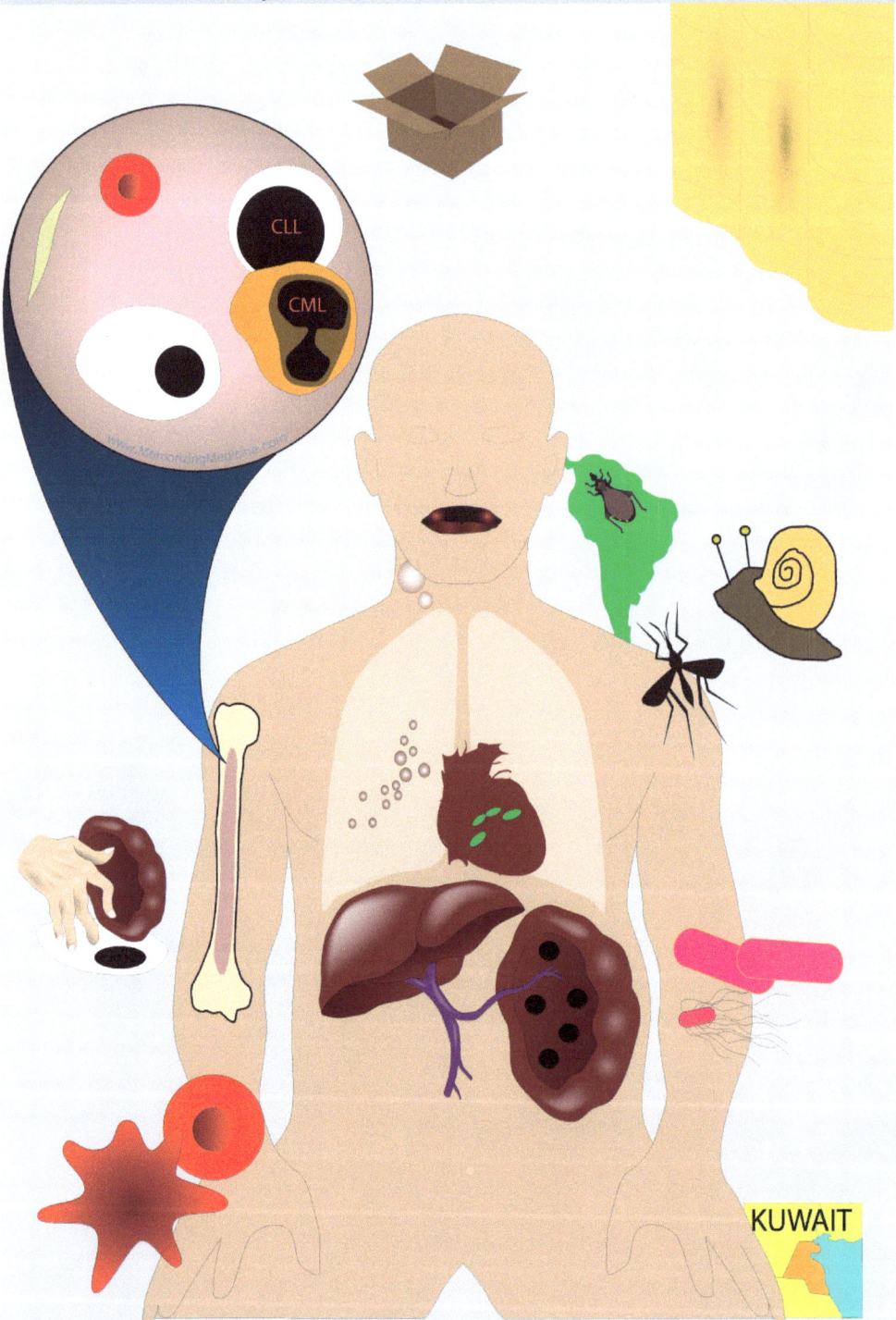

CLL

CML

KUWAIT

© 2010 John Kenneth Dickson

The causes of **Hepatomegaly**

These 3 causes of hepatomegaly also cause cirrhosis. *Note that cirrhosis does not always lead to hepatomegaly.*

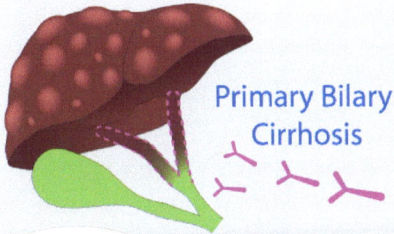

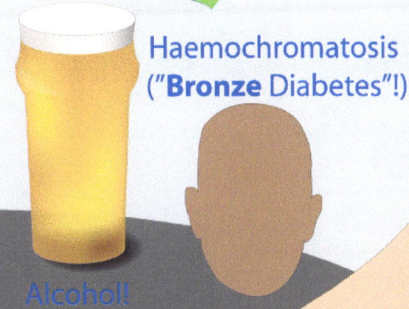

Amyloidosis
"Beta-pleated sheets"!

Infections
eg, Hepatitis & Glandular fever

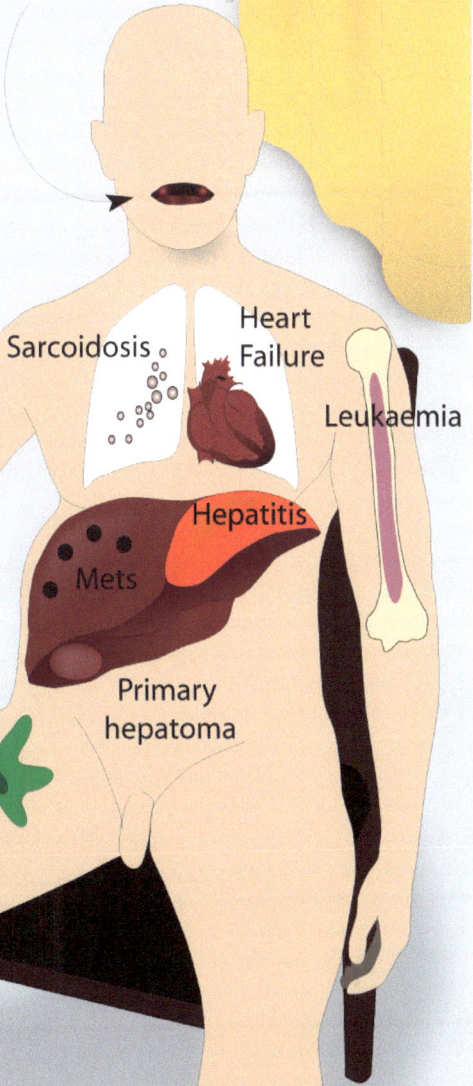

Primary Bilary Cirrhosis

Haemochromatosis ("**Bronze** Diabetes"!)

Sarcoidosis

Heart Failure

Leukaemia

Alcohol!

Hepatitis

Mets

Primary hepatoma

Storage diseases
eg Gaucher's, Neimann-Pick, Tay-Sach's.

Amoebic cysts!

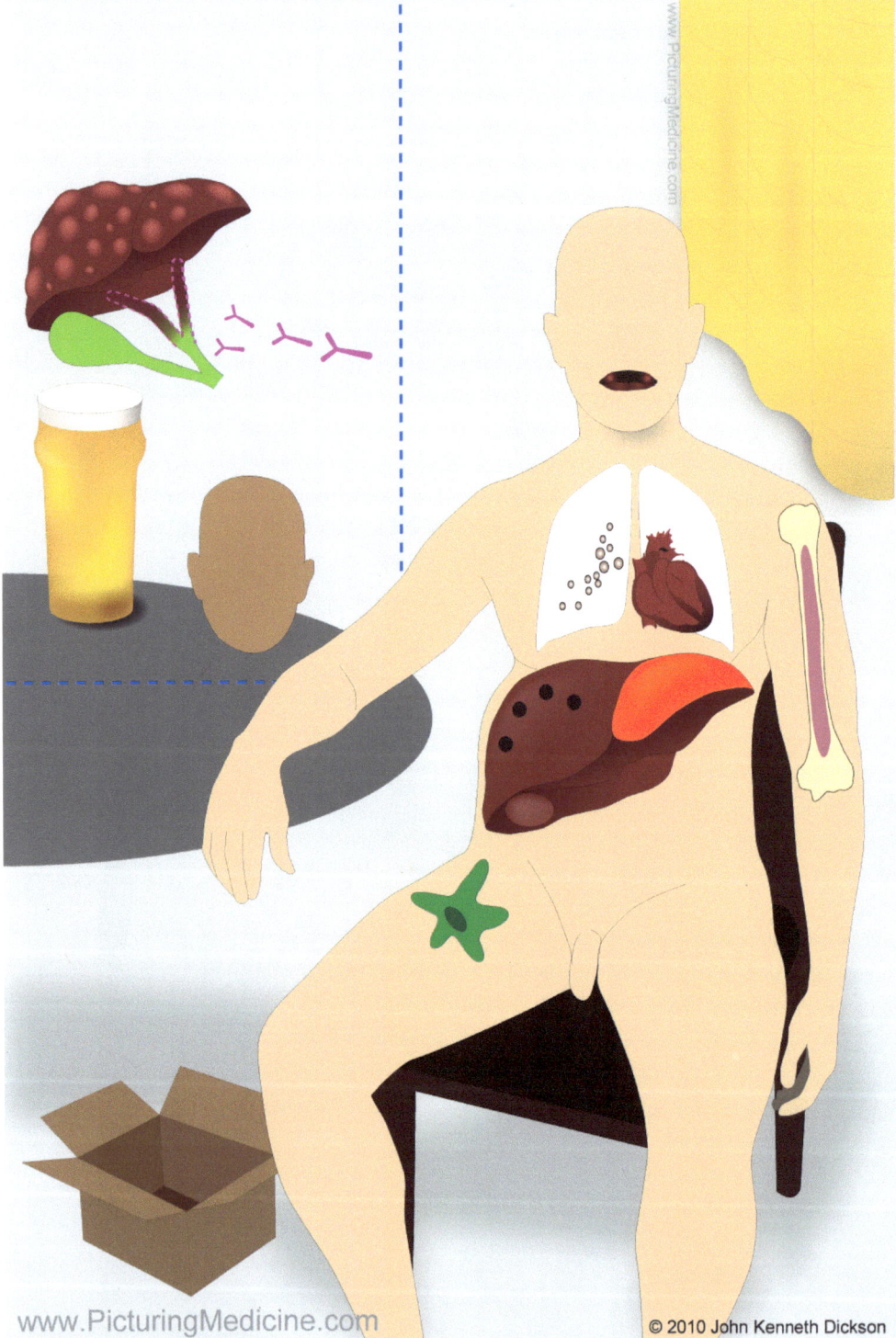

The causes of **Cirrhosis**

Haemachromatosis
"Bronze Diabetes"
(Iron accumulates in the
tissues, inclulding the liver)

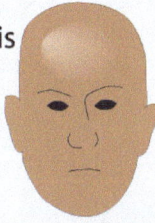

alpha-1 Antitrypsin
Deficiency
(deposition of excessive
abnormal A1AT protein
in liver cells occurs).

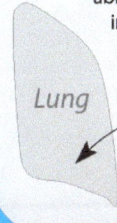

Chronic Viral
Hepatitis

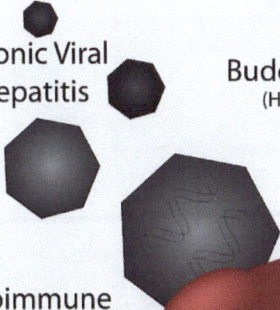

Budd Chiari Syndrome
(Hepatic Vein Thrombosis)

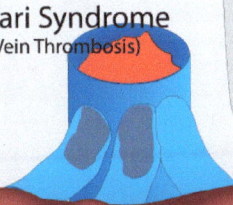

Lung

α-1

Autoimmune
Liver Disease

Wilson's
Disease
(Copper
accumulates
in tissues).

Primary Sclerosing
Cholangitis
(Progressive inflammation
and scarring of the bile
ducts of the liver).

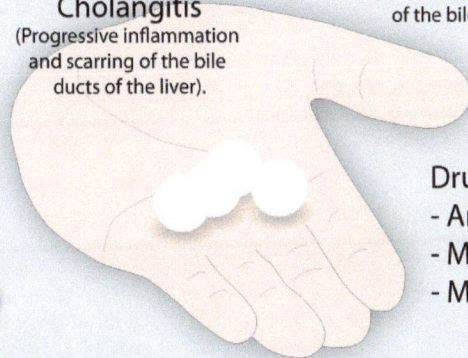

Primary Biliary
Cirrhosis
(Autoimmune attack
of the bile ducts)

Alcohol

Drugs
- Amiodarone
- Methyldopa
- Methotrexate

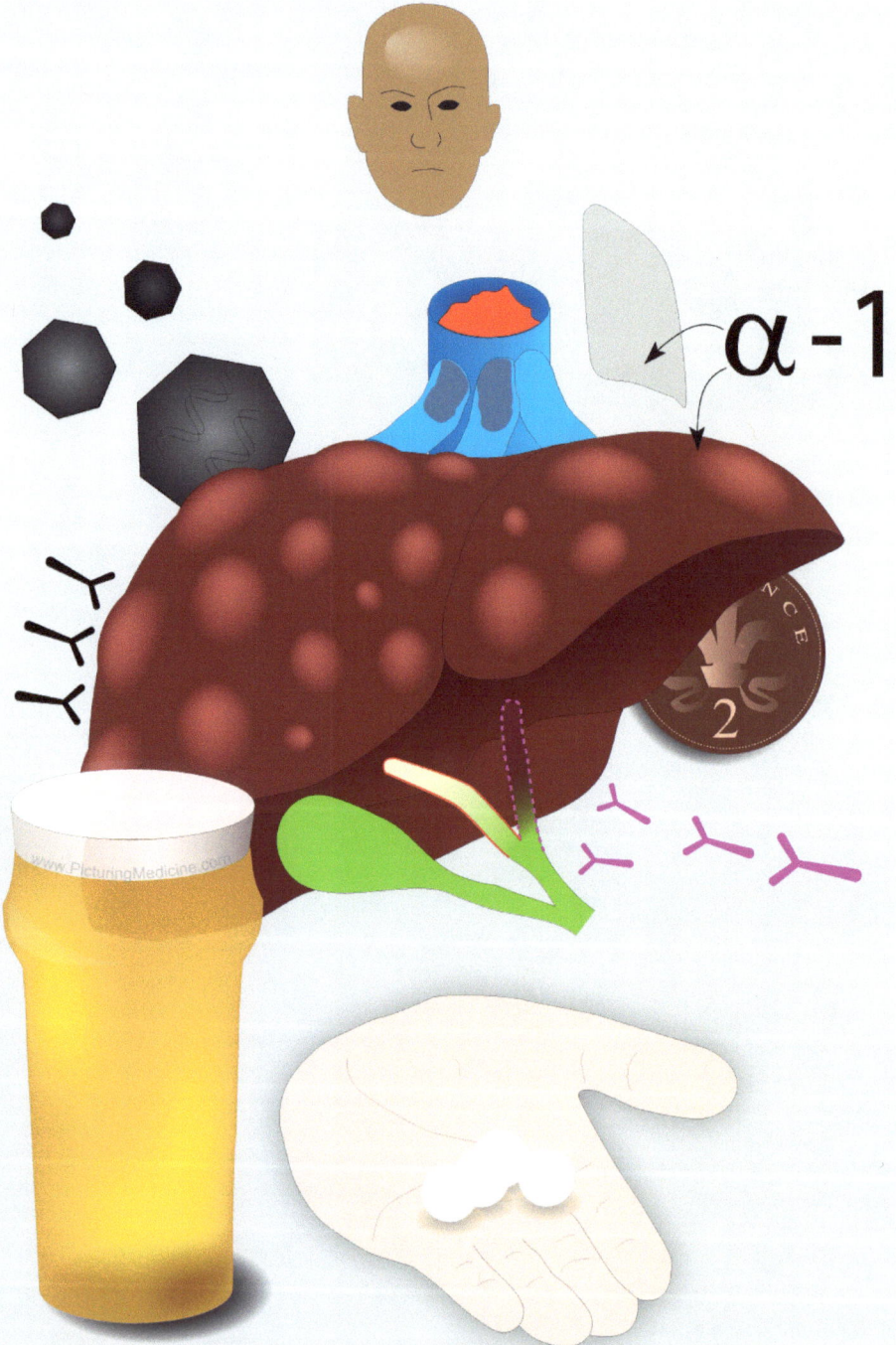

α-1

The causes of **Jaundice**

Uncongugated

Pre-hepatic

Haemolysis

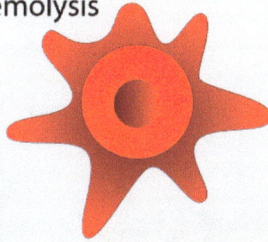

In neontaes
- Physiological Jaundice
- Crigler-Najjar Syndrome
- Haemolysis (Rhesus incompatability!)

Gilbert Syndrome
Reduced glucuronyltransferase activity.
Occurs in 5% of the population.
Mild jaundice requiring no treatment.

Kernicterus
(Brain damage)
can occur!

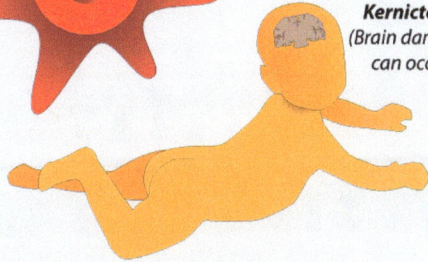

www.PicturingMedicine.com

Congugated

Hepatic

With Hepatic jaundice, some cholestasis can occur occasionally leading to pale stools.

Post-hepatic

Most commonly caused by **gallstones** and **pancreatic cancer**. See separate diagram from more extensive differential diagnosis.

Septicaemia

Right Heart Failure

Acute Hepatitis
(Viral/Autoimmune)

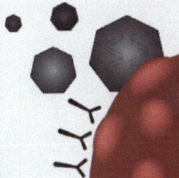

Cirrhosis
(Many causes)

Liver Mets

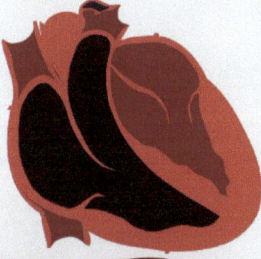

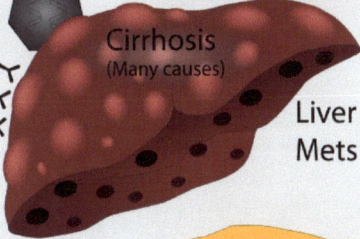

Hepatotoxicity
(Drug-related hepatitis)
- **Paracetamol**
- Anti-TB
- Statins
- Na Valproate

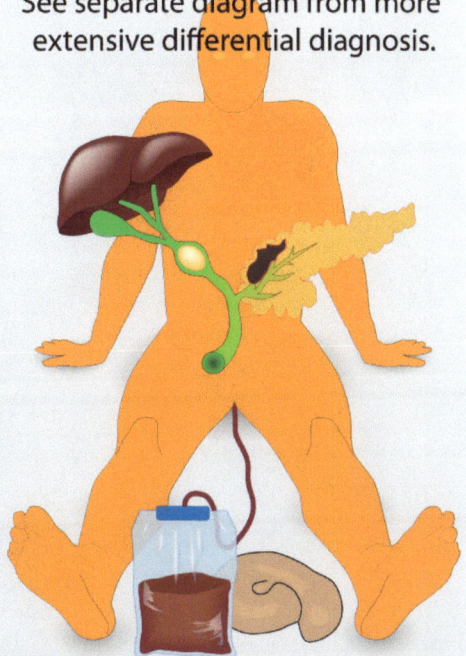

Dark Urine   *Pale Stools*

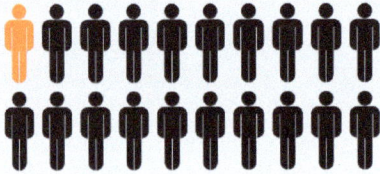

www.PicturingMedicine.com

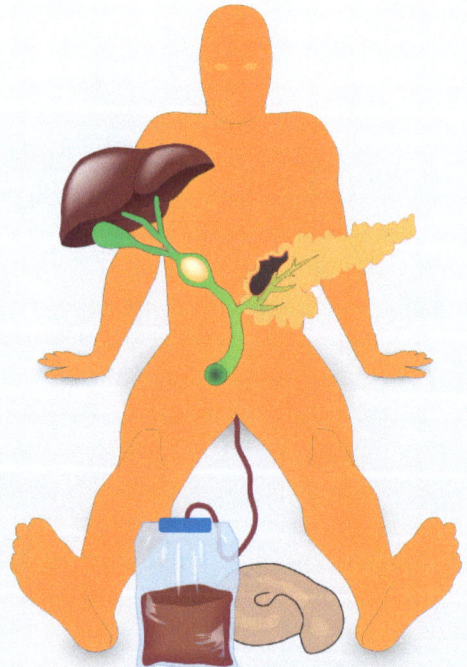

Dark Urine   Pale Stools

The causes of **Delirium**

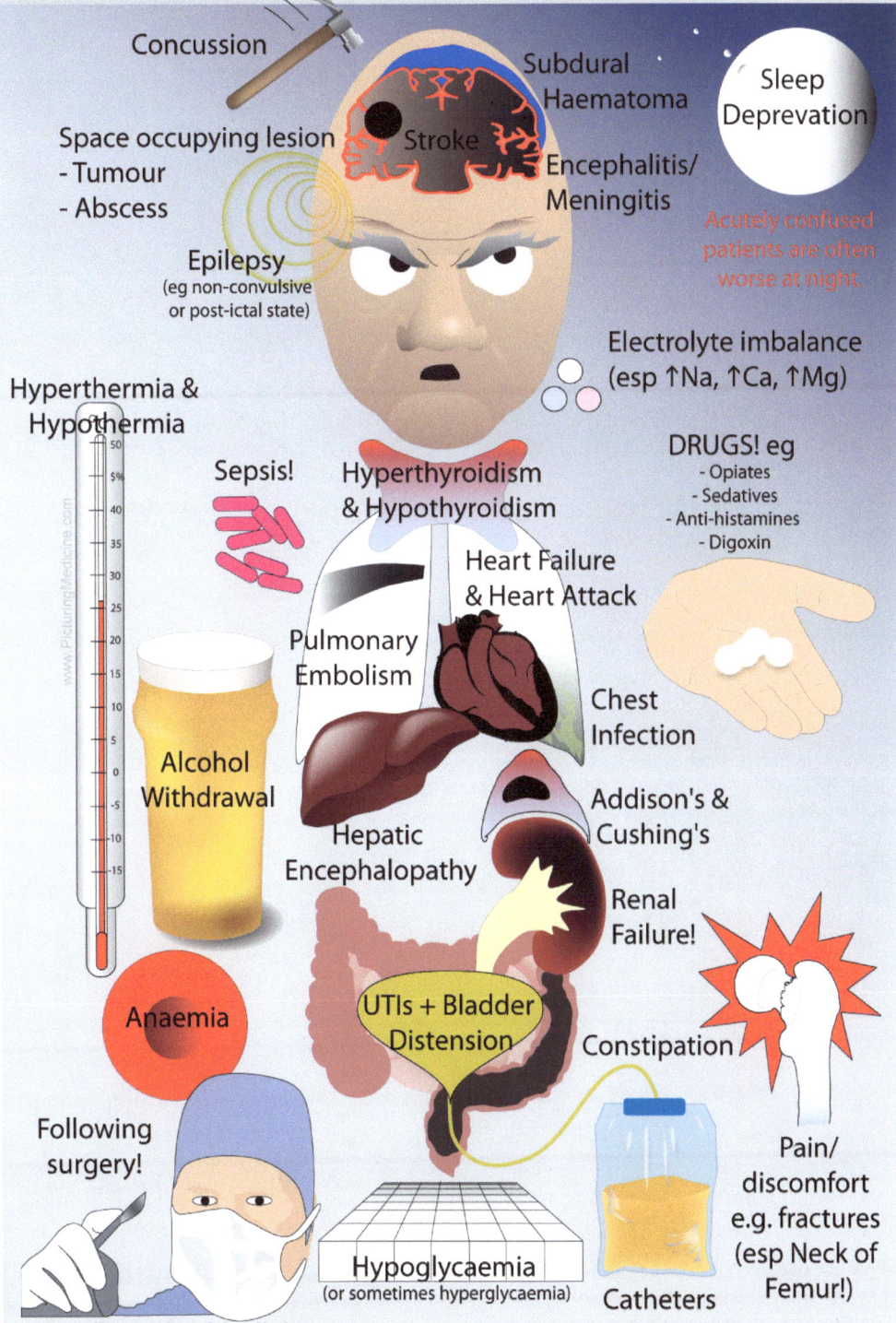

Concussion

Subdural Haematoma

Stroke

Sleep Deprevation

Space occupying lesion
- Tumour
- Abscess

Encephalitis/ Meningitis

Acutely confused patients are often worse at night.

Epilepsy
(eg non-convulsive or post-ictal state)

Electrolyte imbalance
(esp ↑Na, ↑Ca, ↑Mg)

Hyperthermia & Hypothermia

Sepsis!

Hyperthyroidism & Hypothyroidism

DRUGS! eg
- Opiates
- Sedatives
- Anti-histamines
- Digoxin

Heart Failure & Heart Attack

Pulmonary Embolism

Alcohol Withdrawal

Chest Infection

Addison's & Cushing's

Hepatic Encephalopathy

Renal Failure!

Anaemia

UTIs + Bladder Distension

Constipation

Following surgery!

Pain/ discomfort e.g. fractures (esp Neck of Femur!)

Hypoglycaemia
(or sometimes hyperglycaemia)

Catheters

© 2010 John Kenneth Dickson

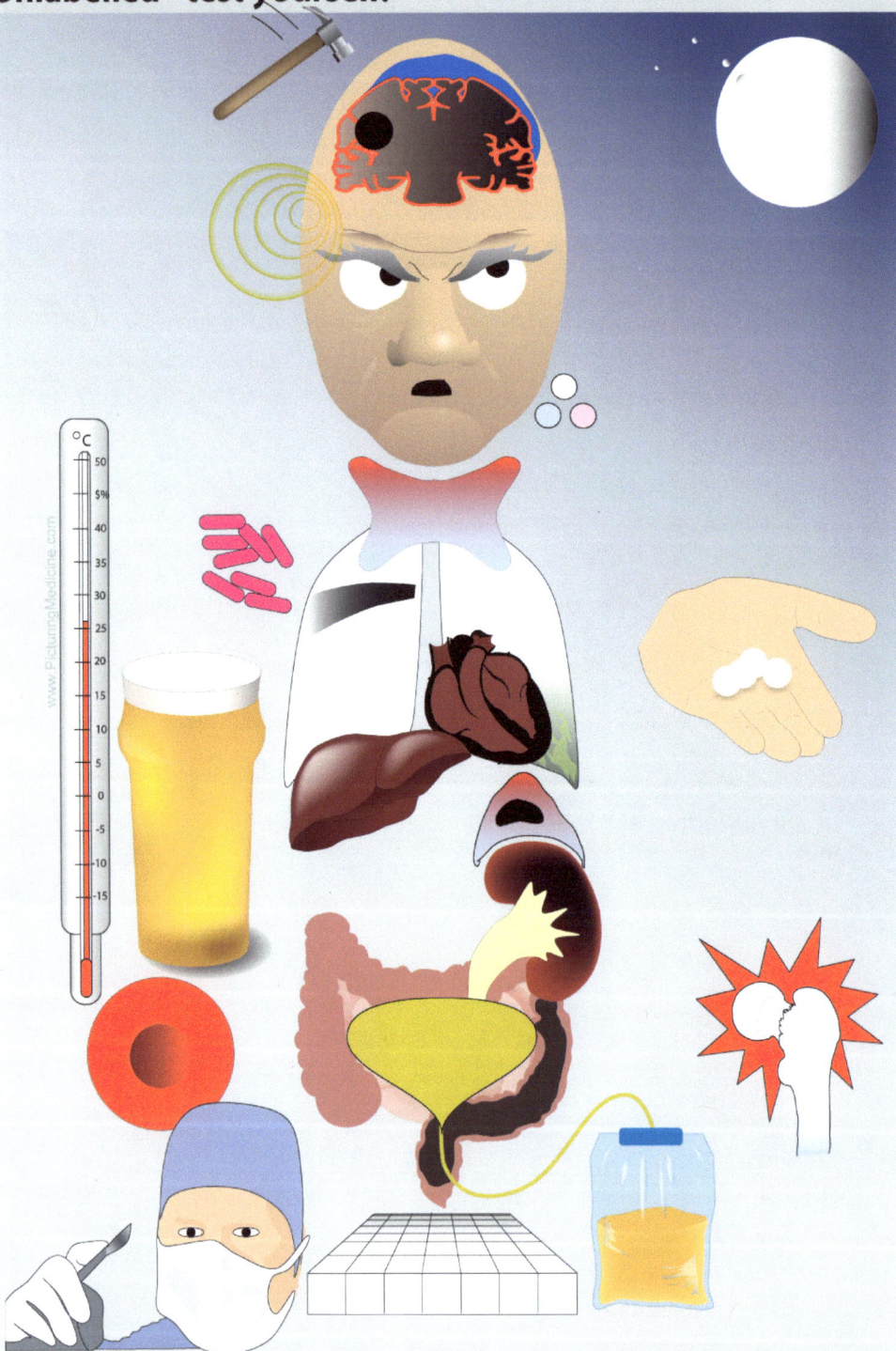

The causes of **Peripheral Neuropathy** Classic "Glove & Stocking Distribution".

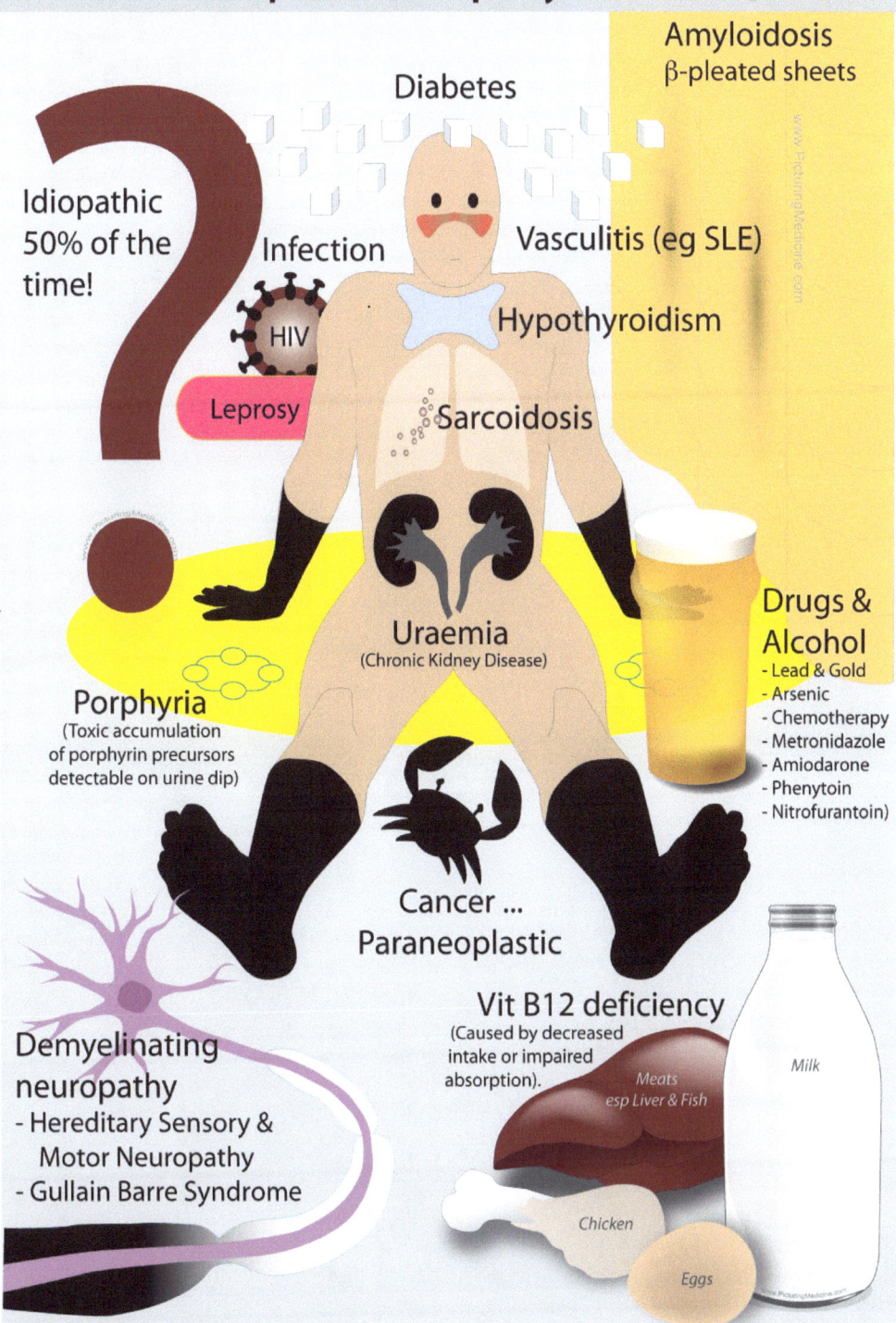

Amyloidosis
β-pleated sheets

Diabetes

Idiopathic
50% of the
time!

Infection

HIV

Leprosy

Vasculitis (eg SLE)

Hypothyroidism

Sarcoidosis

Uraemia
(Chronic Kidney Disease)

Porphyria
(Toxic accumulation
of porphyrin precursors
detectable on urine dip)

Drugs &
Alcohol
- Lead & Gold
- Arsenic
- Chemotherapy
- Metronidazole
- Amiodarone
- Phenytoin
- Nitrofurantoin)

Cancer ...
Paraneoplastic

Vit B12 deficiency
(Caused by decreased
intake or impaired
absorption).

Meats
esp Liver & Fish

Milk

Demyelinating
neuropathy
- Hereditary Sensory &
  Motor Neuropathy
- Gullain Barre Syndrome

Chicken

Eggs

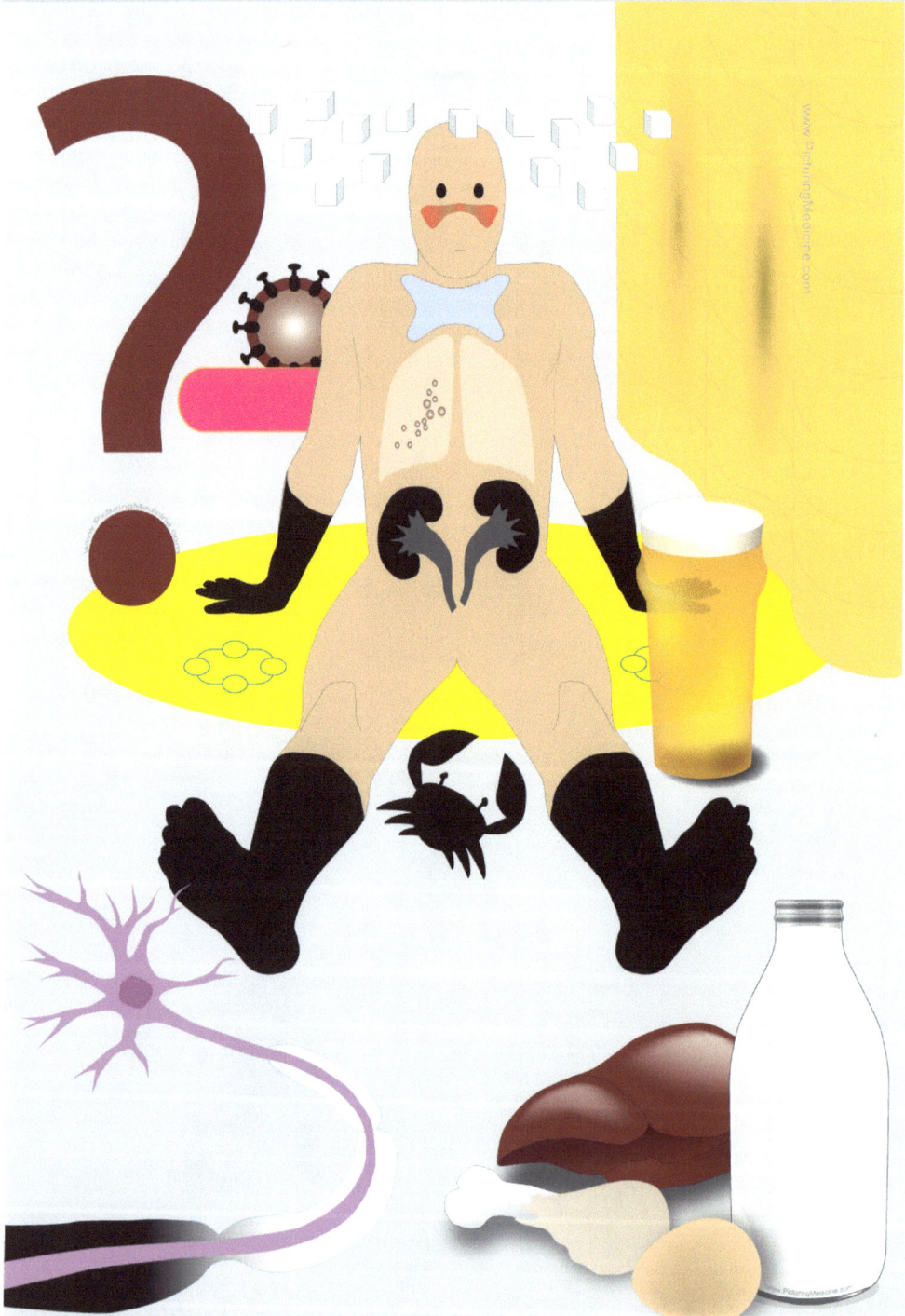

© 2010 John Kenneth Dickson

The causes of **Falls in the Elderly**

Common causes.

Stroke or Transient Ischaemic Attack

Visual Impairment!

Hazards in the Home

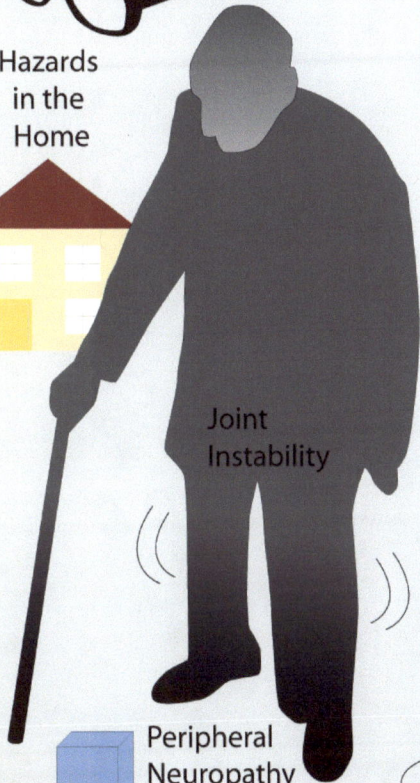

Joint Instability

Peripheral Neuropathy

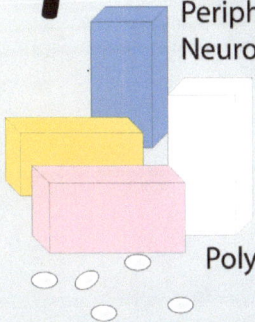

Polypharmacy

Less common causes.

Dementia & Delirium

Syncope

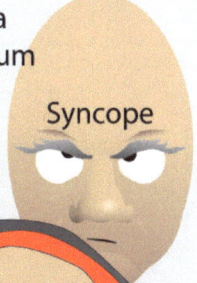

Depression

Subdural Heamatoma

Vestibular Disorders

Seizures

Orthostatic Hypotension

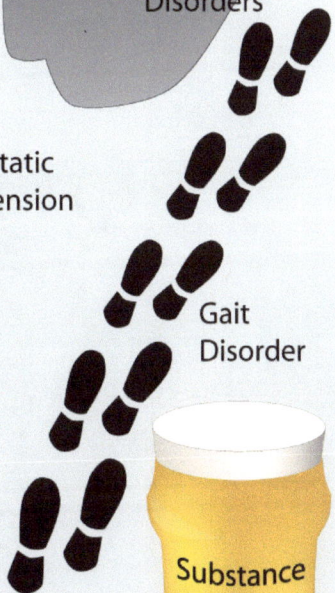

Gait Disorder

Opiates

Opium Poppy!

Substance Abuse
- Alcohol?
- Pain Killers?

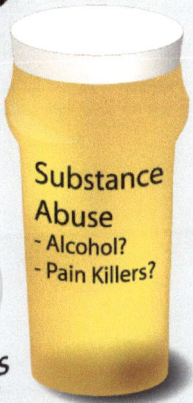

Sedatives
Minor Tranquilizers

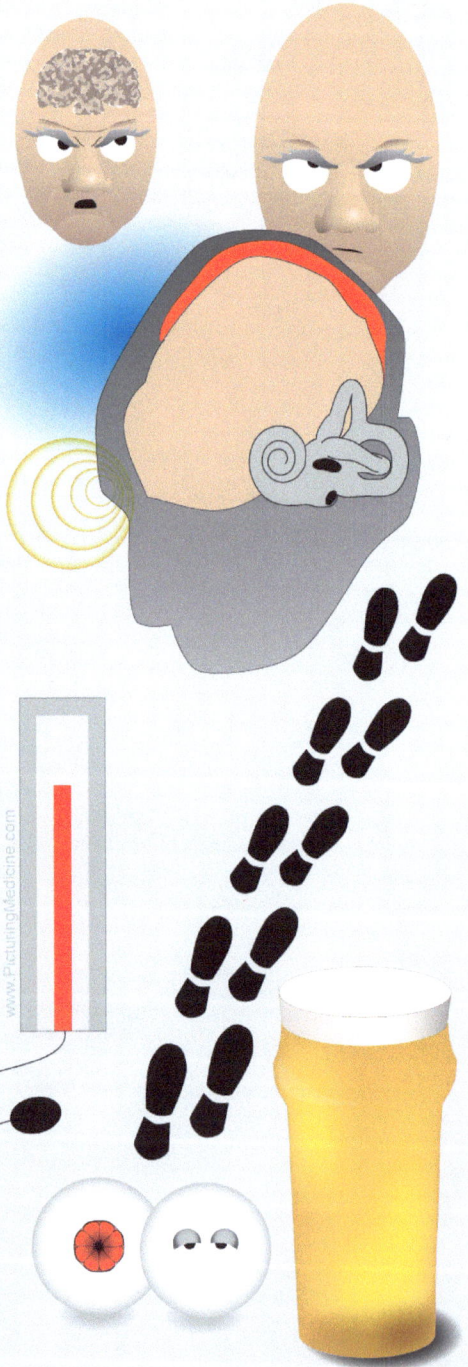

The causes of **Clubbing**

Nail clubbing
A deformity of the nails that is associated with a number of diseases.

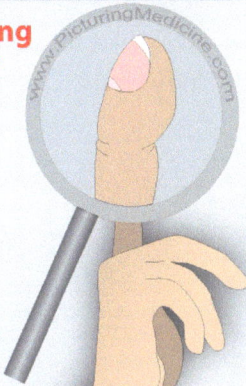

Malignant Mesothelioma
(A form of Lung CA associated with inhalation of asbestos fibres)

Hyperthyroidism
(thyroid acropachy)

Vascular anomalies of the affected arm such as an axillary artery aneurysm (causes unilateral clubbing)

Suppurative lung disease

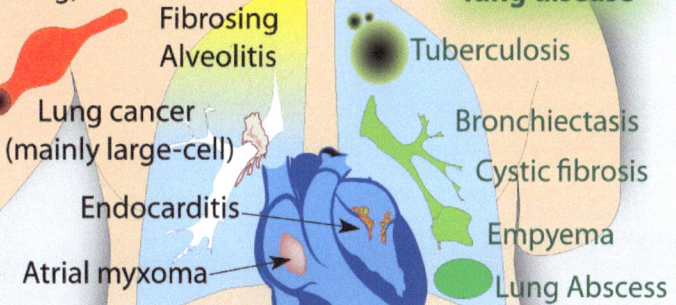

Fibrosing Alveolitis

Tuberculosis

Lung cancer (mainly large-cell)

Bronchiectasis

Cystic fibrosis

Endocarditis

Atrial myxoma

Empyema

Lung Abscess

Congenital cyanotic heart disease

Cirrhosis, esp in primary biliary cirrhosis

GI Lymphoma

Familial and racial clubbing and "pseudoclubbing" (people of African descent often have what appears to be clubbing)

Malabsorption (eg Coeliac)

IBD (Inflammatory Bowel Disease)
- Crohn's disease
- Ulcerative colitis

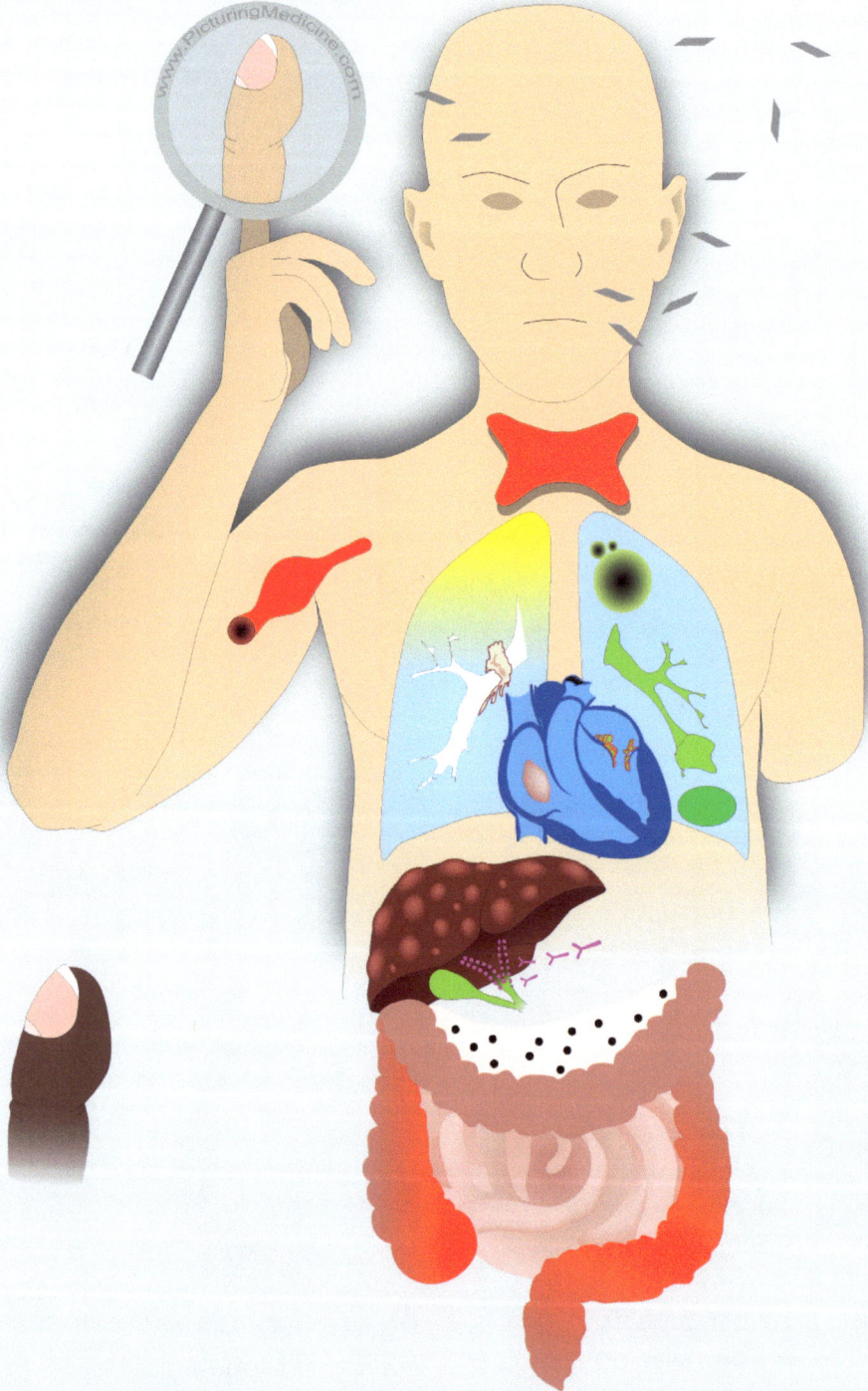

The causes of **Fatigue - "tired all the time"**

More common causes ...

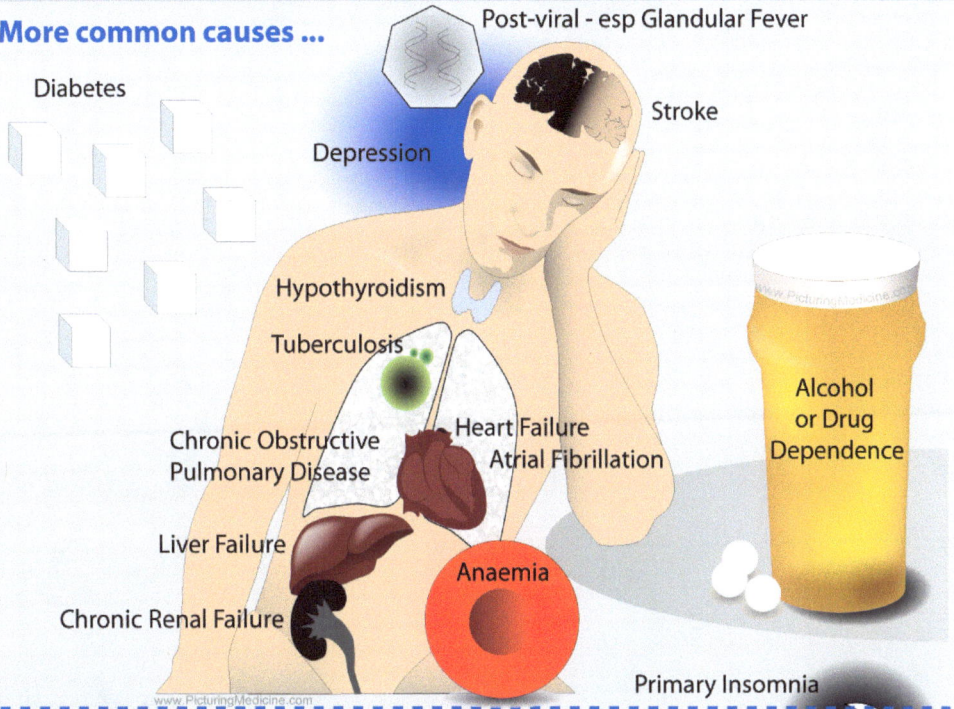

Post-viral - esp Glandular Fever

Diabetes

Stroke

Depression

Hypothyroidism

Tuberculosis

Chronic Obstructive
Pulmonary Disease

Heart Failure
Atrial Fibrillation

Alcohol
or Drug
Dependence

Liver Failure

Anaemia

Chronic Renal Failure

Primary Insomnia

Less common causes ...

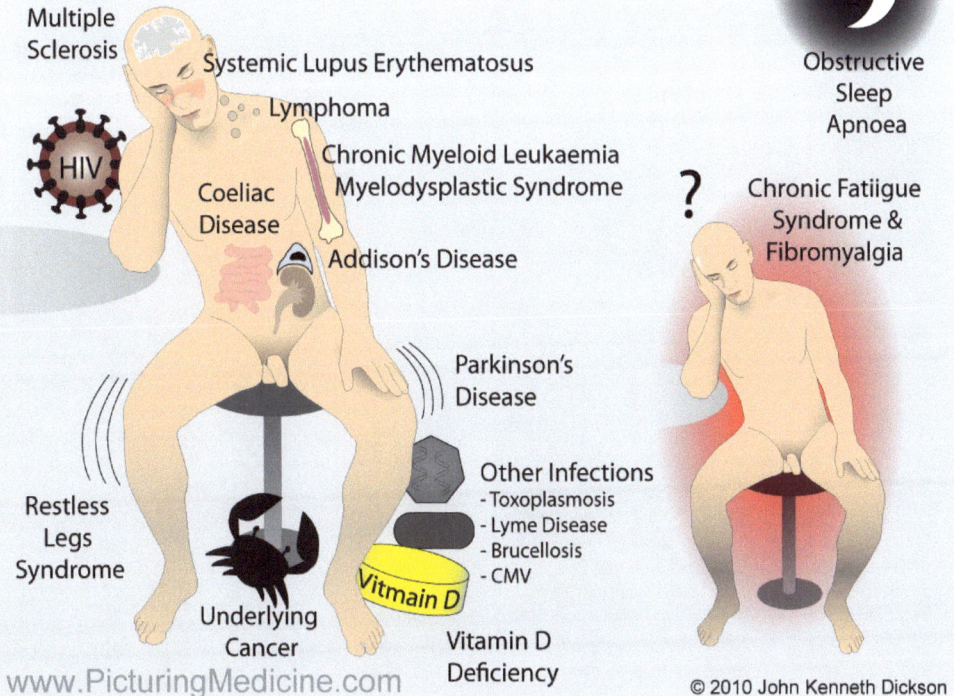

Multiple
Sclerosis

Systemic Lupus Erythematosus

Obstructive
Sleep
Apnoea

Lymphoma

HIV

Chronic Myeloid Leukaemia
Myelodysplastic Syndrome

?

Chronic Fatiigue
Syndrome &
Fibromyalgia

Coeliac
Disease

Addison's Disease

Parkinson's
Disease

Other Infections
- Toxoplasmosis
- Lyme Disease
- Brucellosis
- CMV

Restless
Legs
Syndrome

Vitmain D

Underlying
Cancer

Vitamin D
Deficiency

www.PicturingMedicine.com

© 2010 John Kenneth Dickson

71

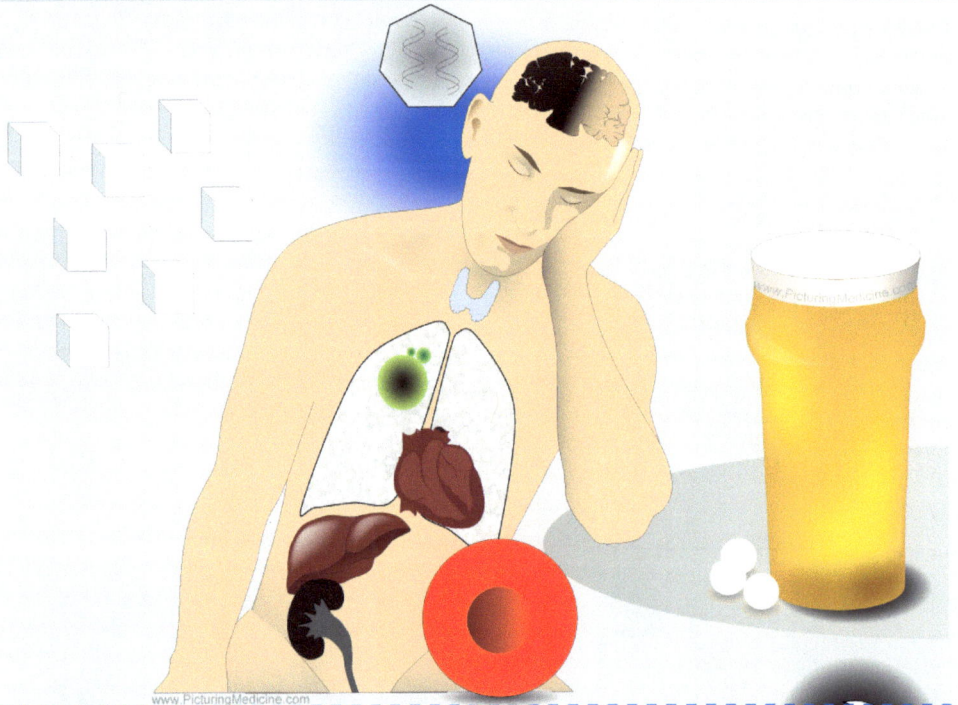

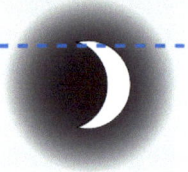

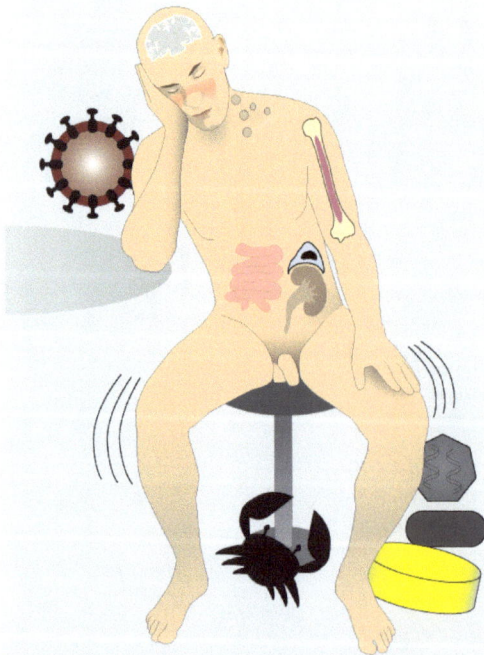

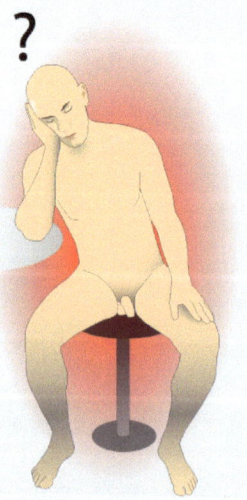

72

The causes of **Vertigo** A sensation that the room is spinning. Grouped here according to duration of symptoms.

Lasting seconds

Post-concussion

Perilymphatic Fistula

Benign Positional Vertigo

Postural Hypotension

Epilepsy

Transient Iscahemic Attack

Arrhythmias

Lasting Hours

Migraine

Meniere's Disease
Swelling within the endolymphatc space.

Alcohol related!

www.PicturingMedicine.com

Lasting Days

Vestibular Neuritis
(aka Acute Labyrinthitis)
Inflammation of the vestibular nerve.
Can sometimes last weeks or even months!
Usually a viral infection.

Stroke

Lasting Months

Acoustic Neuroma

Multiple Sclerosis

Cerebellar Tumour

Medications
- Aminoglycosides
- Benzodiazepines
- Valproate

Unlabelled. Test yourself!

Lasting seconds 🕐

Lasting Hours 🕐🕐🕐

Lasting Days | M | T | W | T | F | S | S |

Lasting Months

M	T	W	T	F	S	S
M	T	W	T	F	S	S
M	T	W	T	F	S	S
M	T	W	T	F	S	S

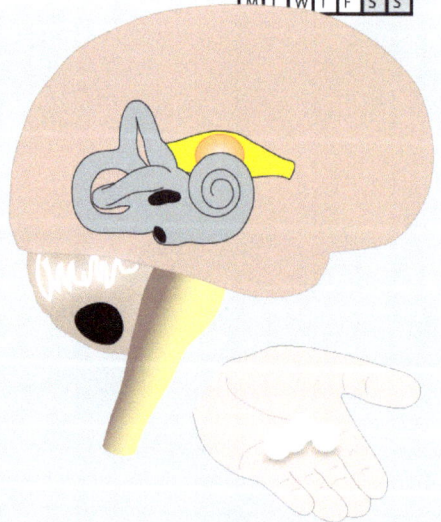

www.ingramcontent.com/pod-product-compliance
Lightning Source LLC
Chambersburg PA
CBHW041312210326

41599CB00003B/81